Study Guide

# PRECALCULUS

**James DeFranza**
St. Lawrence University

**J. Douglas Faires**
Youngstown State University

Brooks/Cole Publishing Company

I(T)P® An International Thomson Publishing Company

Pacific Grove • Albany • Belmont • Bonn • Boston • Cincinnati • Detroit
Johannesburg • London • Madrid • Melbourne • Mexico City
New York • Paris • Singapore • Tokyo • Toronto • Washington

Assistant Editor: *Beth Wilbur*
Cover Design: *Sharon Kinghan*
Cover Photo: *David Bishop/Phototake, NYC*
Editorial Associate: *Nancy Conti*
Marketing Manager: *Jill Downey*
Production Editor: *Mary Vezilich*
Printing and Binding: *Malloy Lithographing*
Disk Duplication: *InserTec*

COPYRIGHT© 1997 by Brooks/Cole Publishing Company
A division of International Thomson Publishing Inc.

 The ITP logo is a registered trademark under license.

*For more information, contact:*

BROOKS/COLE PUBLISHING COMPANY
511 Forest Lodge Road
Pacific Grove, CA 93950
USA

International Thomson Editores
Seneca 53
Col. Polanco
11560 México, D. F., México

International Thomson Publishing Europe
Berkshire House 168-173
High Holborn
London WC1V 7AA
England

International Thomson Publishing Japan
Hirakawacho Kyowa Building, 3F
2-2-1 Hirakawacho
Chiyoda-ku, Tokyo 102
Japan

Thomas Nelson Australia
102 Dodds Street
South Melbourne, 3205
Victoria, Australia

International Thomson Publishing Asia
221 Henderson Road
#05-10 Henderson Building
Singapore 0315

Nelson Canada
1120 Birchmount Road
Scarborough, Ontario
Canada M1K 5G4

International Thomson Publishing GmbH
Königswinterer Strasse 418
53227 Bonn
Germany

All rights reserved. No part of this work may be reproduced, stored in a retrieval system, or transcribed, in any form or by any means—electronic, mechanical, photocopying, recording, or otherwise—without the prior written permission of the publisher, Brooks/Cole Publishing Company, Pacific Grove, California 93950.

Printed in the United States of America

5  4  3  2  1

ISBN 0-534-34583-2

# PREFACE

This Student Study Guide for Precalculus by Faires and DeFranza has been prepared to help students who would like to have access to more review material than the textbook provides. The textbook has been designed to weave the algebra review material into the sections as it is needed, rather than present it as a block of material at the beginning of the book. In this Guide you will find extensive algebra review in the early sections, and additional algebra, geometry, and trigonometry topics thoughout.

The Guide also contains supplemental examples for each of the sections in the textbook, in addition to solutions to many of the odd exercises in the book. The answers to all the odd exercises can be found in the back of the book, but in this Guide you will find the worked out solution to those exercises that we feel are particularly important. The topics presented here should give you an indication of the material in the book that must be mastered if you are to succeed in Precalculus and in Calculus.

Two copies of a placement examination of the type that is used at Youngstown State University can be found in the Appendix at the end of the Guide, or can be down loaded from the WWW site

http://www.cis.ysu.edu/a_s/mathematics/JDFaires/precalculus

We suggest that you work one of these examinations before you take your Precalculus course and the other after you complete the course. If you score 16 or higher on the 40 question examination we would expect that you are sufficiently prepared to take a Precalculus course based on this book. A score of 28 would indicate to us that you are well-prepared for a University Calculus sequence.

We hope this Guide helps in your study of Precalculus and the Calculus courses you will be taking. If you have any suggestions for improvements that can be incorporated into future editions of the book or into this supplement, we would be most grateful for your comments. We can be most easily contacted by electronic mail at the addresses listed below.

St. Lawrence University                                   James DeFranza
                                                          defranza@vm.stlawu.edu

Youngstown State University                               J. Douglas Faires
                                                          faires@math.ysu.edu

November 12, 1996

# Contents

**1 Functions**   1
- 1.1 Introduction . . . . . . . . . . . . . . . . . . . . . . . . . . . 1
- 1.2 The Real Line . . . . . . . . . . . . . . . . . . . . . . . . . . 1
  - 1.2.1 Real Numbers . . . . . . . . . . . . . . . . . . . . . . 1
  - 1.2.2 Set and Interval Notation . . . . . . . . . . . . . . . 3
  - 1.2.3 Exponents and Radicals . . . . . . . . . . . . . . . . 6
  - 1.2.4 Special Products and Factoring . . . . . . . . . . . . 8
  - 1.2.5 Solving Equations and Inequalities . . . . . . . . . . 12
- 1.3 The Cartesian Plane . . . . . . . . . . . . . . . . . . . . . . 20
  - 1.3.1 Describing Regions in the Plane . . . . . . . . . . . 20
  - 1.3.2 The Distance Formula . . . . . . . . . . . . . . . . . 21
  - 1.3.3 The Equation of a Circle . . . . . . . . . . . . . . . 22
- 1.4 The Graph of an Equation . . . . . . . . . . . . . . . . . . . 24
  - 1.4.1 Plotting Points . . . . . . . . . . . . . . . . . . . . . 25
  - 1.4.2 Symmetry . . . . . . . . . . . . . . . . . . . . . . . . 26
- 1.5 Using Technology to Graph Equations . . . . . . . . . . . . 29
- 1.6 Functions . . . . . . . . . . . . . . . . . . . . . . . . . . . . 34
  - 1.6.1 Evaluation of Functions . . . . . . . . . . . . . . . . 34
  - 1.6.2 Domain and Range . . . . . . . . . . . . . . . . . . . 35
  - 1.6.3 Vertical and Horizontal Line Tests . . . . . . . . . . 37
  - 1.6.4 Difference Quotients . . . . . . . . . . . . . . . . . . 38
  - 1.6.5 Odd and Even Functions . . . . . . . . . . . . . . . 40
  - 1.6.6 Applications . . . . . . . . . . . . . . . . . . . . . . 42
- 1.7 Linear Functions . . . . . . . . . . . . . . . . . . . . . . . . 44
  - 1.7.1 The Slope of a Line . . . . . . . . . . . . . . . . . . 44
  - 1.7.2 Point-Slope Equation of a Line . . . . . . . . . . . . 46
  - 1.7.3 Slope-Intercept Equation of a Line . . . . . . . . . . 48

|   |      | 1.7.4 Parallel and Perpendicular Lines | 49 |
|---|------|---|---|
|   |      | 1.7.5 Applications | 51 |
|   | 1.8  | Quadratic Functions | 53 |
|   |      | 1.8.1 Completing the Square | 54 |
|   |      | 1.8.2 Horizontal and Vertical Shifts | 54 |
|   |      | 1.8.3 Horizontal and Vertical Scaling and Reflection | 56 |
|   |      | 1.8.4 Applications | 60 |
|   | 1.9  | Other Common Functions | 63 |
|   |      | 1.9.1 The Absolute Value Function | 63 |
|   |      | 1.9.2 The Square Root Function | 67 |
|   |      | 1.9.3 The Greatest Integer Function | 69 |
|   | 1.10 | Arithmetic Combinations of Functions | 71 |
|   |      | 1.10.1 The Reciprocal Graphing Technique | 76 |
|   | 1.11 | Composition of Functions | 79 |
|   | 1.12 | Inverse Functions | 84 |
|   |      | 1.12.1 One-to-One Functions | 84 |
|   |      | 1.12.2 Process for Finding an Inverse Function | 86 |

## 2 Algebraic Functions — 91

|   |     |   |   |
|---|-----|---|---|
| 2.1 | Introduction | | 91 |
| 2.2 | Polynomial Functions | | 91 |
|     | 2.2.1 | Graphing Polynomial Functions | 92 |
|     | 2.2.2 | Zeros and End Behavior in Sketching Polynomials | 95 |
|     | 2.2.3 | Applications | 103 |
| 2.3 | Finding Factors and Zeros of Polynomials | | 104 |
|     | 2.3.1 | Division of Polynomials | 105 |
|     | 2.3.2 | The Factor Theorem | 106 |
|     | 2.3.3 | The Rational Zero Test | 107 |
|     | 2.3.4 | Descarte's Rule of Signs | 111 |
| 2.4 | Rational Functions | | 112 |
|     | 2.4.1 | Domains and Ranges of Rational Functions | 112 |
|     | 2.4.2 | Horizontal and Vertical Asymptotes | 114 |
|     | 2.4.3 | Slant Asymptotes | 119 |
|     | 2.4.4 | Application to Optimization | 121 |
| 2.5 | Other Algebraic Functions | | 123 |
|     | 2.5.1 | Power Functions | 123 |
| 2.6 | Complex Roots of Polynomials | | 127 |
|     | 2.6.1 | Arithmetic Operations on Complex Numbers | 127 |

      2.6.2    Complex Zeros of Polynomials . . . . . . . . . . . 129

# 3 Trigonometric Functions           133
  3.1  Introduction . . . . . . . . . . . . . . . . . . . . . . . . . . . 133
  3.2  The Sine and Cosine Functions . . . . . . . . . . . . . . . . . 133
        3.2.1    The Unit Circle . . . . . . . . . . . . . . . . . . 133
        3.2.2    Terminal Points . . . . . . . . . . . . . . . . . . 135
        3.2.3    Reference Number . . . . . . . . . . . . . . . . 138
        3.2.4    Sine and Cosine . . . . . . . . . . . . . . . . . . 139
  3.3  Graphs of the Sine and Cosine Functions . . . . . . . . . . . 142
  3.4  Other Trigonometric Functions . . . . . . . . . . . . . . . . 147
        3.4.1    Other Graphs . . . . . . . . . . . . . . . . . . . 153
  3.5  Trigonometric Identities . . . . . . . . . . . . . . . . . . . . 153
        3.5.1    Fundamental Identities . . . . . . . . . . . . . . 153
        3.5.2    Addition and Subtraction Identities . . . . . . . . . . 157
        3.5.3    Double Angle Formulas . . . . . . . . . . . . . . 159
        3.5.4    Half Angle Formulas . . . . . . . . . . . . . . . 161
        3.5.5    Product-to-Sum and Sum-to-Product Identities . . . . 162
        3.5.6    Solving Trigonometric Equations . . . . . . . . . . . 163
  3.6  Right Triangle Trigonometry . . . . . . . . . . . . . . . . . . 164
        3.6.1    Angle Measure: Radians and Degrees . . . . . . . . . 164
        3.6.2    Trigonometric Functions of an Angle in a Right Triangle 165
        3.6.3    Applications . . . . . . . . . . . . . . . . . . . . 168
  3.7  Inverse Trigonometric Functions . . . . . . . . . . . . . . . . 170
        3.7.1    Inverse Sine . . . . . . . . . . . . . . . . . . . . 171
        3.7.2    Summary of the Inverse Trigonometric Functions . . . 172
  3.8  Applications of Trigonometric Functions . . . . . . . . . . . 178
        3.8.1    Law of Cosines . . . . . . . . . . . . . . . . . . 178
        3.8.2    Law of Sines . . . . . . . . . . . . . . . . . . . 180
        3.8.3    Heron's Formula . . . . . . . . . . . . . . . . . 183

# 4 Exponential and Logarithm Functions       185
  4.1  Introduction . . . . . . . . . . . . . . . . . . . . . . . . . . . 185
  4.2  The Natural Exponential Function . . . . . . . . . . . . . . . 185
        4.2.1    Graphs of Exponential Functions . . . . . . . . . . . 186
        4.2.2    The Natural Exponential Function . . . . . . . . . . 190
        4.2.3    Solving Equations Involving $e^x$ . . . . . . . . . . 194
        4.2.4    Compound Interest . . . . . . . . . . . . . . . . 194

    4.3  Logarithm Functions . . . . . . . . . . . . . . . . . . . . 196
           4.3.1  Evaluation of Logarithms . . . . . . . . . . . . . . 196
           4.3.2  Arithmetic Properties of Logarithms . . . . . . . . . 199
           4.3.3  Graphs of Logarithm Functions . . . . . . . . . . . 202
    4.4  Exponential Growth and Decay . . . . . . . . . . . . . . . 205

# 5  Conic Sections, Polar Coordinates, and Parametic Equations  213
    5.1  Introduction . . . . . . . . . . . . . . . . . . . . . . . . . 213
    5.2  Parabolas . . . . . . . . . . . . . . . . . . . . . . . . . . 213
           5.2.1  Standard Position Parabolas . . . . . . . . . . . . . 214
           5.2.2  Applications . . . . . . . . . . . . . . . . . . . . . 222
    5.3  Ellipses . . . . . . . . . . . . . . . . . . . . . . . . . . . 226
           5.3.1  Standard Position Ellipses . . . . . . . . . . . . . . 226
    5.4  Hyperbolas . . . . . . . . . . . . . . . . . . . . . . . . . 230
           5.4.1  Standard Position Hyperbolas . . . . . . . . . . . . 231
           5.4.2  Applications . . . . . . . . . . . . . . . . . . . . . 236
    5.5  Polar Coordinates . . . . . . . . . . . . . . . . . . . . . . 237
           5.5.1  Polar Coordinates To Rectangular Coordinates . . . . 239
           5.5.2  Rectangular Coordinates To Polar Coordinates . . . . 242
           5.5.3  Graphs of Polar Equations . . . . . . . . . . . . . . 244
           5.5.4  Intersection of Polar Curves . . . . . . . . . . . . . 246
    5.6  Conic Sections in Polar Coordinates . . . . . . . . . . . . 247
           5.6.1  Application . . . . . . . . . . . . . . . . . . . . . 251
    5.7  Parametric Equations . . . . . . . . . . . . . . . . . . . . 252
           5.7.1  Using Graphing Devices . . . . . . . . . . . . . . . 259
    5.8  Rotation of Axes . . . . . . . . . . . . . . . . . . . . . . 260

# Chapter 1

# Functions

## 1.1 Introduction

All branches of mathematics are layered vertically, with the most basic material at the bottom of the stack and the most sophisticated at the top. In order to master the most difficult material you must first become versed in the most basic and work up slowly toward the top. This is certainly the case with Algebra, PreCalculus, and Calculus.

In this chapter we review topics from elementary algebra that will be used freely in the material that follows as part of PreCalculus. The first topic is the most basic of all, the number system that is the foundation for most of our work in PreCalculus.

## 1.2 The Real Line

### 1.2.1 Real Numbers

The collection of *real numbers* consists of several different kinds of numbers. The *natural numbers* $N$, consist of all positive whole numbers, 1, 2, 3, 4, .... The *integers* $Z$ consist of the positive and negative whole numbers and 0, that is, $\ldots -3, -2, -1, 0, 1, 2, 3, \ldots$. The *rational numbers* $Q$, consist of all fractions of the form $\frac{p}{q}$, where $p$ and $q$ are integers with $q \neq 0$. Finally, the *irrational numbers*, consist of all real numbers that cannot be represented as a rational numbers, for example, $\sqrt{2}$ or $\pi$.

An important way to distinguish between rational and irrational numbers is through the decimal expansions of these numbers. Rational numbers have decimal expansions that have a repeating pattern and irrational numbers are those with no repeating pattern.

**Example 1** *Classify each number as a natural, integer, rational, or irrational number.*

(a) 192.237  (b) −2.132132132...  (c) 2.01001000100001...  (d) −3

Solution:

(a) Rational since $192.237 = 192\frac{237}{1000} = \frac{192237}{1000}$.

(b) Rational since the decimal expansion has the pattern 132 repeated indefinitely.

(c) Irrational since the number can not be any finite sequence of digits that repeats indefinitely.

(d) Rational and integer, but not a natural number.

■

A letter used to represent an arbitrary real number is called a *variable*. A letter used to represent a fixed value, which does not change, is called a *constant*. An *algebraic expression* is any combination of variables and constants formed using a finite number of the operations of addition, subtraction, multiplication, division, raising to a power, and taking roots. Examples are,

$$ax^2, \quad \frac{x^3 + \sqrt{2x-1}}{x^5 + x^3 - 1}, \quad \text{and} \quad x^2yz^4 - 3xyz^2 + 2x^3y - 6.$$

The first and third examples are called polynomials and will be considered in more detail later.

The basic rules for combining real numbers are simple but important to recognize when manipulating algebraic expressions.

Commutative Laws: $x + y = y + x$ and $xy = yx$
Associative Laws: $(x + y) + z = x + (y + z)$ and $(xy)z = x(yz)$
Distributive Law: $x(y + z) = xy + xz$

**Example 2** *Simplify the algebraic expression using the properties for combining real numbers.*

(a) $2(x - 3) - 3(2 + 5x)$  (b) $(2x^2 + 3xy)(x - 2xy^2)$

Solution: (a) First use the Distributive Law to multiply the 2 and the $-3$ through the parentheses, being careful to multiply both terms in the second parentheses by $-3$. Then combining like terms gives

$$2(x-3) - 3(2+5x) = 2x - 6 - 6 - 15x = -13x - 12.$$

(b) Similarly,

$$\begin{aligned}(2x^2 + 3xy)(x - 2xy^2) &= (2x^2 + 3xy)x - (2x^2 + 3xy)2xy^2 \\ &= 2x^3 + 3x^2y - 4x^3y^2 - 6x^2y^3\end{aligned}$$

We used the Distributive Law twice, with the end result that each term in the first parentheses is multiplied by each term in the second.
∎

### 1.2.2 Set and Interval Notation

A *set* is a collection of objects, called *elements*. Sets are usually denoted using capital letters and elements are usually denoted using lower case letters. There are three standard ways of representing a set:

(i) listing all the elements in brackets $\{\}$;

(ii) listing a few of the elements that give the pattern followed by ...; and,

(iii) describing a property or properties that all elements satisfy in the form $\{x \mid x \text{ satisfies the property } P\}$.

The set with no elements, called the *empty set*, is denoted $\phi$. The two basic operations on sets are *union* ($\cup$) and *intersection* ($\cap$). If $A$ and $B$ are sets, then $A \cup B$ is the set of all elements from either $A$ or $B$, and the intersection $A \cap B$ is the set of all elements common to both sets.

**Example 3** *Determine the intersection and union of the given sets.*

(a) $A = \{-5, -2, 0, 1, 2, 3, 6, 8, 13\}$, $B = \{-3, -2, -1, 5, 7, 9, 11, 13, 15\}$

(b) $A = \{x \mid -2 < x < 10\}$, $B = \{x \mid x \geq 3\}$

Solution:
(a) We have
$A \cup B = \{-5, -3, -2, -1, 0, 1, 2, 3, 5, 6, 7, 8, 9, 11, 13, 15\}$
$A \cap B = \{-2, 13\}$

(b) The union is the set of all real numbers to the right of $-2$, not including $-2$, so $A \cup B = \{x \mid x > -2\}$. Since 3 lies between $-2$ and 10, the intersection is $A \cap B = \{x \mid 3 \leq x < 10\}$.

# CHAPTER 1. FUNCTIONS

■

An *interval* is a set of real numbers between two given real numbers where the endpoints may or may not be included.

**Example 4** *Rewrite the sets using interval notation. Sketch the intervals on a real line.*

(a) $\{x \mid -1 < x < 2\}$  (b) $\{x \mid 2 \leq x < 5\}$  (c) $\{x \mid -2 < x \leq 0\}$
(d) $\{x \mid -5 \leq x \leq -2\}$  (e) $\{x \mid x < 3\}$  (f) $\{x \mid x \leq -1\}$
(g) $\{x \mid x > -2\}$  (h) $\{x \mid x \geq 0\}$

Solution:
(a) $\{x \mid -1 < x < 2\} = (-1, 2)$

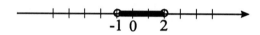

(b) $\{x \mid 2 \leq x < 5\} = [2, 5)$

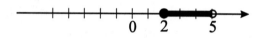

(c) $\{x \mid -2 < x \leq 0\} = (-2, 0]$

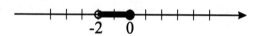

(d) $\{x \mid -5 \leq x \leq -2\} = [-5, -2]$

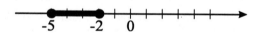

(e) $\{x \mid x < 3\} = (-\infty, 3)$

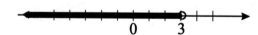

(f) $\{x \mid x \leq -1\} = (-\infty, -1]$

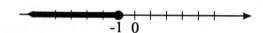

(g) $\{x \mid x > -2\} = (-2, \infty)$

(h) $\{x \mid x \geq 0\} = [0, \infty)$

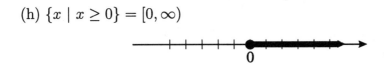

Notice that when an endpoint is not included in the interval, a round parenthesis, ( or ), is used. When the endpoint is included, a square bracket, [ or ], is used. Since $\infty$ is only a symbol, not a real number, only round parentheses are placed next to $\infty$ and $-\infty$.

■

The length of an interval is the distance between two points on the real line. The *absolute value* of a real number is the distance from the number to the origin, 0. The absolute value of a number $x$ is defined by

$$|x| = \begin{cases} x, & \text{if } x \geq 0 \\ -x, & \text{if } x < 0. \end{cases}$$

**Example 5** *Determine the exact value of the absolute value.*
(a) $|3.5|$ (b) $|-3|$ (c) $|\sqrt{2} - 1|$ (d) $|1 - \sqrt{3}|$

Solution:
(a) $|3.5| = 3.5$
(b) $|-3| = -(-3) = 3$
(c) $|\sqrt{2} - 1| = \sqrt{2} - 1$, since $\sqrt{2} > 1$.
(d) $|1 - \sqrt{3}| = -(1 - \sqrt{3}) = \sqrt{3} - 1$ since $\sqrt{3} > 1$, so $1 - \sqrt{3} < 0$.
■

The distance between two real numbers $a$ and $b$ is $|a - b| = |b - a|$.

**Example 6** *Determine the distance between the two real numbers.*
(a) 2 and 12.7  (b) $-3$ and 5

Solution:
(a) $|2 - 12.7| = |-10.7| = 10.7$
(b) $|5 - (-3)| = |-3 - 5| = 8$

Notice that the order of subtraction does not matter as long as the computations are done inside the absolute value.

■

## 1.2.3 Exponents and Radicals

Listed are the basic definitions and properties.

$$x^n = \underbrace{x \cdot x \cdot x \cdots x}_{n\text{-times}} \qquad x^{-n} = \frac{1}{x^n} \qquad x^{\frac{1}{n}} = \sqrt[n]{x}$$

$$x^m x^n = x^{m+n} \qquad (xy)^n = x^n y^n \qquad (x^m)^n = x^{mn}$$

$$\frac{x^m}{x^n} = x^{m-n} \qquad \left(\frac{x}{y}\right)^n = \frac{x^n}{y^n}$$

$$\sqrt[n]{xy} = \sqrt[n]{x}\sqrt[n]{y} \qquad \sqrt[n]{\frac{x}{y}} = \frac{\sqrt[n]{x}}{\sqrt[n]{y}} \qquad x^{\frac{m}{n}} = \sqrt[n]{x^m} = \left(\sqrt[n]{x}\right)^m$$

**Example 7** *Simplify each of the expressions. Do not use negative exponents in the final answer.*
(a) $x^3 x^4$ (b) $x^2 x^{-9}$ (c) $\frac{x^5}{x^2}$ (d) $\frac{x^2}{x^7}$

Solution:
(a) $x^3 x^4 = x^{3+4} = x^7$
(b) $x^2 x^{-9} = x^{2-9} = x^{-7} = \frac{1}{x^7}$
(c) $\frac{x^5}{x^2} = x^{5-2} = x^3$
(d) $\frac{x^2}{x^7} = x^{2-7} = x^{-5} = \frac{1}{x^5}$

A shortcut in problems like (c) and (d) is to subtract the smaller exponent from the larger, placing the result in the numerator or denominator depending on which has the term with the larger exponent.
∎

**Example 8** *Simplify each of the expressions.*
(a) $(2x^3 y^2)(x^4 y^3)^3$ (b) $\left(\frac{x^2 y}{xy}\right)^3 \left(\frac{xy^5}{x^6 y^2}\right)^2$ (c) $\left(\frac{2x^{-2} y^3}{3x^4 y}\right)^{-2}$

Solution:
(a) First bring the outer exponent of 3 inside the second parentheses, by multiplying each inside exponent by 3, and then combine the exponents on like terms. This gives
$$(2x^3 y^2)(x^4 y^3)^3 = (2x^3 y^2)(x^{12} y^9) = 2x^{15} y^{11}.$$

## 1.2. THE REAL LINE

(b) $\left(\dfrac{x^2y}{xy}\right)^3 \left(\dfrac{xy^5}{x^6y^2}\right)^2 = (x)^3 \left(\dfrac{y^3}{x^5}\right)^2 = x^3 \dfrac{y^6}{x^{10}} = \dfrac{y^6}{x^7}$

(c) $\left(\dfrac{2x^{-2}y^3}{3x^4y}\right)^{-2} = \left(\dfrac{2y^2}{3x^6}\right)^{-2} = \dfrac{2^{-2}y^{-4}}{3^{-2}x^{-12}} = \dfrac{9x^{12}}{4y^4}$

In parts (b) and (c) it is slightly easier to simplify inside the parentheses first and then apply the outside exponent, as we did in the solution. Try to solve the problem another way.

∎

**Example 9** *Rationalize each of the fractions involving radicals.*

(a) $\dfrac{1}{\sqrt{3}}$  (b) $\sqrt{\dfrac{3}{5}}$  (c) $\dfrac{\sqrt{3}-1}{\sqrt{3}+1}$

Solution: *Rationalizing* a fraction involving radicals refers to eliminating all radicals from the denominator of the expression, by multiplying by a suitable fraction equaling 1. This was done originally to simplify calculation that was performed by hand. It is much easier to divide radical approximations by integers, for example, than it is to divide integers by radical approximations. While this is no longer an important application because of the wide-spread availability of calculators, it is still a useful procedure.

(a) $\dfrac{1}{\sqrt{3}} = \dfrac{1}{\sqrt{3}}\dfrac{\sqrt{3}}{\sqrt{3}} = \dfrac{\sqrt{3}}{3}$

(b) $\sqrt{\dfrac{3}{5}} = \dfrac{\sqrt{3}}{\sqrt{5}} = \dfrac{\sqrt{3}\sqrt{5}}{\sqrt{5}\sqrt{5}} = \dfrac{\sqrt{3}\sqrt{5}}{5} = \dfrac{\sqrt{15}}{5}$

(c) $\dfrac{\sqrt{3}-1}{\sqrt{3}+1} = \dfrac{\sqrt{3}-1}{\sqrt{3}+1}\dfrac{\sqrt{3}-1}{\sqrt{3}-1} = \dfrac{3-2\sqrt{3}+1}{3-1} = \dfrac{4-2\sqrt{3}}{2} = 2-\sqrt{3}$

In (c), simply change the sign between the $\sqrt{3}$ and 1 in the denominator from plus to minus and multiply and divide by $\sqrt{3}-1$.

∎

**Example 10** *Simplify each of the expressions.*

(a) $\sqrt[3]{8x^6y^5}$  (b) $\left(\dfrac{9x^{2/3}y^4}{4x^9}\right)^{3/2}$

Solution:

(a) $\sqrt[3]{8x^6y^5} = (8x^6y^5)^{1/3} = 8^{1/3}(x^6)^{1/3}(y^5)^{1/3} = 2x^2y^{5/3}$

(b) $\left(\dfrac{9x^{2/3}y^4}{4x^9}\right)^{3/2} = \dfrac{(\sqrt{9})^3 xy^6}{(\sqrt{4})^3 x^{27/2}} = \dfrac{27xy^6}{8x^{27/2}} = \dfrac{27y^6}{8x^{25/2}}$

In part (b) it is a bit more difficult to first simplify inside the parentheses and then apply the rational exponent 3/2.

∎

### 1.2.4 Special Products and Factoring

The following formulas are very useful for quickly multiplying certain algebraic expressions which arise frequently, and are especially useful when going in the reverse direction of factoring an algebraic expression.

$$(x+y)^2 = x^2 + 2xy + y^2 \qquad (x-y)^2 = x^2 - 2xy + y^2$$

$$(x+y)(x-y) = x^2 - y^2$$

$$(x+y)^3 = x^3 + 3x^2y + 3xy^2 + y^3 \qquad (x-y)^3 = x^3 - 3x^2y + 3xy^2 - y^3$$

$$x^3 + y^3 = (x+y)(x^2 - xy + y^2) \qquad x^3 - y^3 = (x-y)(x^2 + xy + y^2)$$

Notice that $(x+y)^2 \neq x^2 + y^2$, $(x-y)^2 \neq x^2 - y^2$, $(x+y)^3 \neq x^3 + y^3$, and $(x-y)^3 \neq x^3 - y^3$.

**Example 11** *Perform the indicated multiplication.*

(a) $(2x-2y)(2x+2y)$  (b) $(x^2+y^2)^2$  (c) $(3x^3-y)^3$  (d) $(x-\sqrt{x})(x+\sqrt{x})$

Solution: In each part we will apply the appropriate special product formula.

(a) $(2x - 2y)(2x + 2y) = (2x)^2 - (2y)^2 = 4x^2 - 4y^2$

(b) $(x^2 + y^2)^2 = (x^2)^2 + 2x^2y^2 + (y^2)^2 = x^4 + 2x^2y^2 + y^4$

(c)
$$\begin{aligned}(3x^3 - y)^3 &= (3x^3)^3 - 3(3x^3)^2 y + 3(3x^3)y^2 - y^3 \\ &= 27x^9 - 27x^6y + 9x^3y^2 - y^3\end{aligned}$$

(d) $(x - \sqrt{x})(x + \sqrt{x}) = x^2 - (\sqrt{x})^2 = x^2 - x$

The special product formulas are not required to compute the products, although it is easier to use them. For example, in part (b) we could write

$$(x^2 + y^2)^2 = (x^2 + y^2)(x^2 + y^2),$$

multiply each term in the first parenthesis by each term in the second, and then collect like terms to get the desired answer. This would be more difficult

## 1.2. THE REAL LINE

in part (c). To convince yourself of this you might try the problems this alternative way. ■

**Example 12** *Factor each expression.*
(a) $4x^2 - 9y^2$  (b) $x^2 + 4x + 4$  (c) $4x^3 - 20x$  (d) $x^4 + 2x^2 + 1$

Solution: Each problem fits one of the special formulas given.

(a) To get the expression to fit the formula for the difference of two squares, we note that $4x^2 = (2x)^2$ and $9y^2 = (3y)^2$, so

$$4x^2 - 9y^2 = (2x)^2 - (3y)^2 = (2x + 3y)(2x - 3y).$$

(b) $x^2 + 4x + 4 = (x+2)^2$

(c) Always look first for common terms that can be factored out of the expression. This usually simplifies the work. In this case we have

$$4x^3 - 20x = 4x(x^2 - 5) = 4x\left(x^2 - \left(\sqrt{5}\right)^2\right) = 4x\left(x - \sqrt{5}\right)\left(x + \sqrt{5}\right).$$

(d) There are no common factors, but treating the term $x^2$ as the $x$ in the special formulas and the 1 as the $y$ we have

$$x^4 + 2x^2 + 1 = (x^2)^2 + 2(x^2) + 1 = (x^2 + 1)^2.$$

■

Usually the more difficult problems to factor are those that do not fit one of the special formulas. Factoring is needed so often that it is very important to feel comfortable with the process. To factor an expression of the form $ax^2 + bx + c$, it is necessary to choose factors of the numbers $a$ and $c$ and their signs in just the right way.

**Example 13** *Factor each expression.*
(a) $x^2 - x - 2$  (b) $6x^2 + 7x + 2$  (c) $2x^3 - 3x^2 - 2x$

Solution:

(a) The only factor of $a = 1$ is 1, and the only factors of $c = 2$ are 1 and 2. So the only choice for a possible factoring is

$$x^2 - x - 2 = (x \pm 1)(x \pm 2).$$

To get the constant term of $-2$, the signs of 1 and 2 must differ. Since the middle term is $-x$ rather then $x$, the minus sign must be with the 2. So

$$x^2 - x - 2 = (x+1)(x-2).$$

(b) The factors of $a = 6$ are $1, 2, 3, 6$, and the only factors of $c = 2$ are 1 and 2. Since the constant term is positive and the middle term is also positive the signs inside the factors will also be both positive. Both negative would give a positive constant term but a negative middle term. The possibilities are

$$(6x+1)(x+2) = 6x^2 + 13x + 2$$
$$(6x+2)(x+1) = 6x^2 + 8x + 2$$
$$(3x+1)(2x+2) = 6x^2 + 8x + 2$$
$$(3x+2)(2x+1) = 6x^2 + 7x + 2$$

and we see the last one gives the answer.

(c) $2x^3 - 3x^2 - 2x = x(2x^2 - 3x - 2) = x(2x+1)(x-2)$

■

**Example 14** *Simplify each expression.*

(a) $\dfrac{x^2 + 5x + 6}{x^2 - 2x - 8}$ (b) $\dfrac{x^2 + 2x + 1}{x^2 - x - 2} \cdot \dfrac{2x^2 - 5x + 2}{x^2 + 4x + 3}$

(c) $\dfrac{2x^2 + 8x}{x^2 - 3x + 2} \div \dfrac{x^2 + 6x + 8}{x - 1}$

Solution:

(a) When simplifying fractional expressions always try to first factor as much as possible and then cancel any like terms from the numerator and denominator. We have

$$\frac{x^2 + 5x + 6}{x^2 - 2x - 8} = \frac{(x+2)(x+3)}{(x+2)(x-4)} = \frac{x+3}{x-4}.$$

(b) The rule

$$\frac{a}{b} \cdot \frac{c}{d} = \frac{ac}{bd}$$

for multiplying two number fractions is also used for algebraic expressions. So

## 1.2. THE REAL LINE

$$\frac{x^2+2x+1}{x^2-x-2} \cdot \frac{2x^2-5x+2}{x^2+4x+3} = \frac{(x+1)^2}{(x+1)(x-2)} \cdot \frac{(2x-1)(x-2)}{(x+3)(x+1)}$$
$$= \frac{(x+1)^2(2x-1)(x-2)}{(x+1)^2(x+3)(x-2)}$$
$$= \frac{2x-1}{x+3}.$$

(c) The rule
$$\frac{\frac{a}{b}}{\frac{c}{d}} = \frac{a}{b} \cdot \frac{d}{c}$$

for dividing two number fractions is also used for algebraic expressions. So,

$$\frac{2x^2+8x}{x^2-3x+2} \div \frac{x^2+6x+8}{x-1} = \frac{2x^2+8x}{x^2-3x+2} \cdot \frac{x-1}{x^2+6x+8}$$
$$= \frac{2x(x+4)}{(x-1)(x-2)} \cdot \frac{(x-1)}{(x+2)(x+4)}$$
$$= \frac{2x}{(x-2)(x+2)}$$
$$= \frac{2x}{x^2-4}.$$

■

**Example 15** *Simplify each expression.*
(a) $\dfrac{1}{x-2} + \dfrac{4}{x+1}$   (b) $\dfrac{1}{\sqrt{x}} - \dfrac{1}{\sqrt{x+a}}$

Solution:
(a) To add the fractions we need to find a common denominator. The rule
$$\frac{a}{b} + \frac{c}{d} = \frac{ad+cb}{bd}$$
for adding number fractions is used. So

$$\frac{1}{x-2} + \frac{4}{x+1} = \frac{(x+1)+4(x-2)}{(x-2)(x+1)}$$
$$= \frac{5x-7}{x^2-x-2}.$$

(b) Taking the common denominator we have

$$\frac{1}{\sqrt{x}} - \frac{1}{\sqrt{x+a}} = \frac{\sqrt{x+a}-\sqrt{x}}{\sqrt{x}\sqrt{x+a}}.$$

Expressions like these can also be rationalized to eliminate some of the radicals. In this case we eliminate the radical in the numerator. That is,

$$\frac{1}{\sqrt{x}} - \frac{1}{\sqrt{x+a}} = \frac{\sqrt{x+a}-\sqrt{x}}{\sqrt{x}\sqrt{x+a}}$$
$$= \frac{\sqrt{x+a}-\sqrt{x}}{\sqrt{x}\sqrt{x+a}} \cdot \frac{\sqrt{x+a}+\sqrt{x}}{\sqrt{x+a}+\sqrt{x}}$$
$$= \frac{(x+a)-x}{\sqrt{x}\sqrt{x+a}(\sqrt{x+a}+\sqrt{x})}$$
$$= \frac{a}{\sqrt{x}\sqrt{x+a}(\sqrt{x+a}+\sqrt{x})}.$$

In this case we have "rationalized" the numerator rather than the denominator.
∎

## 1.2.5 Solving Equations and Inequalities

Solving equations and inequalities arise as subproblems to many problems encountered in a PreCalculus or Calculus course. To solve an equation or inequality means to find all real numbers that satisfy the given condition. Equations and inequalities are solved by using the basic rules of arithmetic.

**Example 16** *Find all values of $x$ that satisfy the equation.*
(a) $5x - 9 = 6$ (b) $\frac{2}{3}x + 3 = 2$ (c) $\frac{2x}{x+1} + \frac{1}{2} = 3$

## 1.2. THE REAL LINE

Solution:

(a) To solve the equation, first add 9 to *both* sides of the equation and then divide by 5. Then

$$\begin{align*}
5x - 9 &= 6 \\
5x - 9 + 9 &= 6 + 9 \\
5x &= 15 \\
\frac{5x}{5} &= \frac{15}{5} \\
x &= 3.
\end{align*}$$

(b) First subtract 3 from both sides of the equation and then multiply by 3/2.

$$\begin{align*}
\frac{2}{3}x + 3 &= 2 \\
\frac{2}{3}x &= -1 \\
\frac{3}{2}\left(\frac{2}{3}x\right) &= \frac{3}{2}(-1) \\
x &= -\frac{3}{2}
\end{align*}$$

(c) First subtract 1/2 from both sides of the equation and then multiply by $2(x+1)$ assuming $x \neq -1$. Then

$$\begin{align*}
\frac{2x}{x+1} + \frac{1}{2} &= 3 \\
\frac{2x}{x+1} &= \frac{5}{2} \\
2(2x) &= 5(x+1) \\
4x &= 5x + 5 \\
-x &= 5 \\
x &= -5.
\end{align*}$$

■

**Example 17** *Solve each equation.*
(a) $2x^2 + 5x - 3 = 0$ (b) $x^2 + x - 2 = 4$ (c) $x^4 + x^2 - 2 = 0$

Solution:
(a) To solve the equation for $x$ we first factor the left side. So

$$2x^2 + 5x - 3 = 0$$
$$(2x - 1)(x + 3) = 0.$$

The product of two real numbers can be 0 only if at least one of the numbers is 0. So

$$2x - 1 = 0, \quad x + 3 = 0$$
$$x = \frac{1}{2}, \quad x = -3.$$

(b) First rewrite the equation so 0 is on the right side, then factor.

$$x^2 + x - 2 = 4$$
$$x^2 + x - 6 = 0$$
$$(x + 3)(x - 2) = 0$$
$$x = -3, x = 2$$

(c) If we let $u = x^2$, then $x^4 + x^2 - 2 = u^2 + u - 2 = (u + 2)(u - 1)$, so

$$x^4 + x^2 - 2 = 0$$
$$(x^2 + 2)(x^2 - 1) = 0$$
$$x^2 + 2 = 0, \quad x^2 - 1 = 0.$$

Since $x^2 + 2 = 0$ has no solutions, the only solutions are

$$x^2 - 1 = 0$$
$$x^2 = 1$$
$$x = \pm 1.$$

∎

## 1.2. THE REAL LINE

If the solutions to an equation of the form $ax^2 + bx + c = 0$ are rational numbers, with $a \neq 0$, then the expression can be factored and the solutions found as we did above.

In general, the solutions to the *quadratic equation* $ax^2 + bx + c = 0$ with $a \neq 0$, can be found using the *Quadratic formula*. The solutions are given by

$$x = \frac{-b \pm \sqrt{b^2 - 4ac}}{2a}.$$

**Example 18** *Find all solutions to $x^2 + 4x + 2 = 0$ and factor the expression $x^2 + 4x + 2$.*

Solution: Using the Quadratic formula with $a = 1, b = 4$, and $c = 2$, gives

$$\begin{aligned} x &= \frac{-4 \pm \sqrt{16 - 4(1)(2)}}{2} \\ &= \frac{-4 \pm \sqrt{8}}{2} = \frac{-4 \pm 2\sqrt{2}}{2} \\ &= -2 \pm \sqrt{2}. \end{aligned}$$

Once the solutions (called the *roots*) are known, the expression can be factored completely as

$$x^2 + 4x + 2 = \left(x - \left(-2 + \sqrt{2}\right)\right)\left(x - \left(-2 - \sqrt{2}\right)\right).$$

■

In the previous example the solutions are irrational numbers and we needed to use the Quadratic formula. We can not find the factors of the expression using a trial and error approach, as was done when the solutions were rational numbers.

**Example 19** *Find all solutions to the equations.*
  (a) $\sqrt{2x - 1} = 3$  (b) $\sqrt{x - 2} + x = 2$

Solution:
(a) First square both sides and then isolate $x$, giving

$$\begin{aligned} \sqrt{2x - 1} &= 3 \\ 2x - 1 &= 9 \\ x &= 5. \end{aligned}$$

Because we squared the equation we need to check the answer to be sure that we have not introduced and *extraneous* solution. That is, a solution to the final equation that was not a solution to the original equation. In this case we did not since $\sqrt{2(5)-1} = \sqrt{9} = 3$.

(b) In problems like this it is easiest to first isolate the radical term and then square both sides. So

$$\begin{aligned}
\sqrt{x-2} + x &= 2 \\
\sqrt{x-2} &= 2-x \\
x-2 &= (2-x)^2 \\
x-2 &= 4 - 4x + x^2 \\
x^2 - 5x + 6 &= 0 \\
(x-2)(x-3) &= 0 \\
x &= 2, \quad x = 3.
\end{aligned}$$

It is important that you check the answers, since squaring both sides of an equation may introduce extraneous solutions. Checking,

$$\begin{aligned}
\sqrt{3-2} + 3 &= 4 \neq 2 \\
\sqrt{2-2} + 2 &= 2
\end{aligned}$$

and we see $x = 2$ *is* a solution but $x = 3$ *is not* a solution. ■

To solve inequalities a few basic properties are needed.

<u>Properties of Inequalities</u>
1. If $a < b$ and $b < c$, then $a < c$.
2. If $a < c$, then $a + b < c + b$.
3. If $a < b$, and $c > 0$, then $ac < bc$.
4. If $a < b$, and $c < 0$, then $ac > bc$.

**Example 20** *Solve each inequality.*
 (a) $2x - 3 > 2$ (b) $-5x + 2 < 3$

Solution:
(a) First apply Property 2, and then add 3 to both sides to get,

$$\begin{aligned}
2x - 3 + 3 &> 2 + 3 \\
2x &> 5.
\end{aligned}$$

Then use Property 3, and divide both sides by 2, so

$$\frac{2x}{2} > \frac{5}{2}$$
$$x > \frac{5}{2}.$$

In interval notation the solution is $\left(\frac{5}{2}, \infty\right)$.

(b)

$$-5x + 2 < 3$$
$$-5x + 2 - 2 < 3 - 2$$
$$-5x < 1$$
$$\frac{-5x}{-5} > \frac{1}{-5}$$
$$x > -\frac{1}{5}$$

In interval notation the solution can be written $\left(-\frac{5}{2}, \infty\right)$. In the division step, be very careful to reverse the inequality sign since both sides of the inequality are divided by a negative number.

■

**Example 21** *Solve each inequality.*

(a) $x^2 - x - 2 > 0$  (b) $-x^2 - 2x + 3 > 0$  (c) $\dfrac{x^2 + 3x + 2}{x^2 + 1} < 0$

Solution: To solve a quadratic inequality, first try to factor the quadratic expression. In the case of a fractional expression, as in part (c), try to factor numerator and denominator.

(a) Factoring we have

$$x^2 - x - 2 = (x - 2)(x + 1) > 0.$$

In order for the product of two linear factors to be positive, either both factors must be positive or both must be negative. For example,

if $x = 3$, then $(3 - 2)(3 + 1) = 4 > 0$,

if $x = -2$, then $(-2 - 2)(-2 + 1) = (-4)(-1) = 4 > 0$,

but if $x = 1.5$, then $(1.5 - 2)(1.5 + 1) = (-0.5)(2.5) = -1.25 \not> 0$.

The values that make the linear factors 0, in this case $x = 2$ and $x = -1$, separate the real line into three intervals, $(-\infty, -1), (-1, 2)$, and $(2, \infty)$. To solve the inequality, select test values from each interval, substitute the values into the equation, and the inequality will have the same sign for all numbers in the interval.

| Interval | Test value | Test | Inequality |
|---|---|---|---|
| $(-\infty, -1)$ | $-2$ | $(-2 - 2)(-2 + 1) = 4$ | $> 0$ |
| $(-1, 2)$ | $0$ | $(0 - 2)(0 + 1) = -2$ | $< 0$ |
| $(2, \infty)$ | $3$ | $(3 - 2)(3 + 1) = 4$ | $> 0$ |

From the table, the solution to the inequality is $x < -1$ or $x > 2$, that is, $(-\infty, -1) \cup (2, \infty)$.

(b) Since
$$-x^2 - 2x + 3 = -(x^2 + 2x - 3) = -(x + 3)(x - 1)$$
we have the table

| Interval | Test value | Test | Inequality |
|---|---|---|---|
| $(-\infty, -3)$ | $-4$ | $-(-4 + 3)(-4 - 1) = -5$ | $< 0$ |
| $(-3, 1)$ | $0$ | $-(0 + 3)(0 - 1) = 3$ | $> 0$ |
| $(1, \infty)$ | $2$ | $-(2 + 3)(2 - 1) = -5$ | $< 0$ |

The solution is $x$ in the interval $(-3, 1)$.

(c) Since the denominator can not be factored,
$$\frac{x^2 + 3x + 2}{x^2 + 1} = \frac{(x + 2)(x + 1)}{x^2 + 1}.$$
We could proceed as we did in parts (a) and (b), but in examples like this it is easier to first observe that the denominator is always greater than 0, so
$$\frac{x^2 + 3x + 2}{x^2 + 1} < 0 \quad \text{precisely when} \quad (x + 2)(x + 1) < 0.$$

| Interval | Test value | Test | Inequality |
|---|---|---|---|
| $(-\infty, -2)$ | $-3$ | $(-3 + 2)(-3 + 1) = 2$ | $> 0$ |
| $(-2, -1)$ | $-1.5$ | $(-1.5 + 2)(-1.5 + 1) = -0.25$ | $< 0$ |
| $(-1, \infty)$ | $0$ | $(0 + 2)(0 + 1) = 2$ | $> 0$ |

## 1.2. THE REAL LINE

The solution is the interval $(-2, -1)$. ∎

To solve inequalities involving absolute values we use the two facts:

$$|x| < a \text{ means } -a < x < a$$
$$|x| > a \text{ means } x < -a \text{ or } x > a.$$

Often in problems like this an expression involving $x$ is inside the absolute value signs rather than $x$ alone. This is not a concern since the $x$ is just a place holder for any real number. The first statement, for example, could be written as

$$|\Box| < a \text{ means } -a < \Box < a$$

where $\Box$ can be any expression. A similar interpretation can be given to the second statement.

**Example 22** *Solve each inequality.*
(a) $|2x - 1| < 2$  (b) $|3x + 2| > 1$

Solution:
(a)
$$\begin{aligned} |2x - 1| &< 2 \\ -2 < 2x - 1 &< 2 \\ -1 < 2x &< 3 \\ -\tfrac{1}{2} < x &< \tfrac{3}{2} \end{aligned}$$

The solution set is the interval $\left(-\tfrac{1}{2}, \tfrac{3}{2}\right)$.

(b)
$$\begin{aligned} |3x + 2| &> 1 \\ 3x + 2 < -1 \quad &\text{or} \quad 3x + 2 > 1 \\ 3x < -3 \quad &\text{or} \quad 3x > -1 \\ x < -1 \quad &\text{or} \quad x > -\tfrac{1}{3} \end{aligned}$$

In interval notation the solution set is $(-\infty, -1) \cup \left(-\tfrac{1}{3}, \infty\right)$. ∎

## 1.3 The Cartesian Plane

The $xy$-coordinate system is called the coordinate plane or Cartesian plane. Points are specified using an ordered pair $(x, y)$ with the horizontal or $x$-coordinate specified first and the vertical or $y$-coordinate specified second.

### 1.3.1 Describing Regions in the Plane

**Example 23** *(Exercise Set 1.3, Exercise 13) Describe the set of points in the $xy$-plane that satisfy $-1 \leq x \leq 2$.*

Solution: Since the $y$-coordinate can be any real number and the $x$-coordinate is restricted to lie between $-1$ and $2$, the inequality describes all points between the vertical lines determined by $\{(x,y) \mid x = -1\}$ and $\{(x,y) \mid x = 2\}$. For example, $(-1, 0), (-1, 2), (-1, -2), (0, 1), (0, 2)$ are all in the region whereas $(-2, 1)$ and $(3, -1)$ are not in the region.

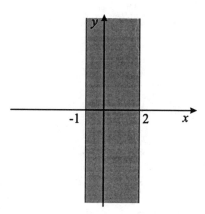

■

**Example 24** *(Exercise Set 1.3, Exercise 17) Describe the set of points in the $xy$-plane satisfying $|x - 1| < 3$ and $|y + 1| < 2$.*

Solution: Rewriting the inequalities we have,

$$\begin{aligned} |x - 1| &< 3 \\ -3 < x - 1 &< 3 \\ -2 < x &< 4 \end{aligned}$$

## 1.3. THE CARTESIAN PLANE

and

$$|y+1| < 2$$
$$-2 < y+1 < 2$$
$$-3 < y < 1.$$

This means the points in the solution set all have $x$-coordinates between $-2$ and $4$, not including the values $-2$ and $4$, and $y$-coordinates between $-3$ and $1$, not including the values $-3$ and $1$. That is, the points lie between, but not on, the vertical lines $x = -2$ and $x = 4$ and between, but not on, the horizontal lines $y = -3$ and $y = 1$ describing a rectangular region as shown in the figure.

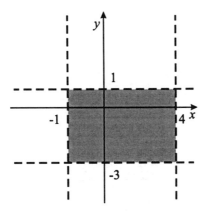

■

### 1.3.2 The Distance Formula

The distance between two points $P(x_1, y_1)$ and $Q(x_2, y_2)$ is given by the formula

$$d(P,Q)) = \sqrt{(x_2 - x_1)^2 + (y_2 - y_1)^2}$$

**Example 25** *(Exercise Set 1.3, Exercise 31) Find the distances between the points $(-1, 4), (-3, -4)$, and $(2, -1)$, and show that they are vertices of a right triangle. At which vertex is the right angle?*

Solution: Let $A = (-1, 4)$, $B = (-3, -4)$, and $C = (2, -1)$. The sketch of the points in the figure indicates the right angle is at $C$. To verify that the triangle is a right triangle we determine the lengths $a, b$, and $c$ and verify that $a^2 + b^2 = c^2$. So

$$\begin{aligned} a &= d(B,C) = \sqrt{(2-(-3))^2 + (-1-(-4))^2} = \sqrt{34} \\ b &= d(A,C) = \sqrt{(2-(-1))^2 + (-1-4)^2} = \sqrt{34} \\ c &= d(A,B) = \sqrt{(-3-(-1))^2 + (-4-4)^2} = \sqrt{68}. \end{aligned}$$

Hence

$$a^2 + b^2 = 34 + 34 = 68$$

and

$$a^2 + b^2 = c^2.$$

The right angle is at $(2, -1)$.

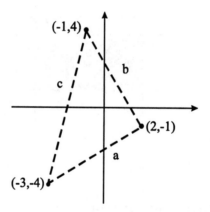

■

### 1.3.3 The Equation of a Circle

A *circle* is a set of points a fixed distance, called the *radius*, from a fixed point, called the *center*. If the radius is $r$ and the fixed point in the plane is $(h, k)$, the distance formula implies that a point $(x, y)$ lies on the circle provided

$$\sqrt{(x-h)^2 + (y-k)^2} = r \quad \text{or} \quad (x-h)^2 + (y-k)^2 = r^2.$$

## 1.3. THE CARTESIAN PLANE

**Example 26** *Find the equation of the circle with center $(-1, 2)$ and radius 2.*

Solution: Substituting into the formula we have the equation of the circle is
$$(x - (-1))^2 + (y - 2)^2 = 4$$
$$(x + 1)^2 + (y - 2)^2 = 4.$$

■

**Example 27** *Find the center and radius of the circle $x^2 - 2x + y^2 + 2y + 1 = 0$.*

Solution: To find the center and radius, rewrite the equation in the general form. To do this, complete the square on the $x$-terms and the $y$-terms of the equation. To complete the square on $x^2 - 2x$, take half the coefficient of the $x$-term, square it, and then add and subtract (so the net change is 0) the value. So
$$\begin{aligned} x^2 - 2x &= x^2 - 2x + \left(\frac{2}{2}\right)^2 - \left(\frac{2}{2}\right)^2 \\ &= x^2 - 2x + 1 - 1 \\ &= (x - 1)^2 - 1. \end{aligned}$$

Doing the same process on the $y$-terms gives
$$\begin{aligned} x^2 - 2x + y^2 + 2y + 1 &= 0 \\ x^2 - 2x + 1 - 1 + y^2 + 2y + 1 - 1 + 1 &= 0 \\ (x - 1)^2 + (y + 1)^2 &= 1 \\ (x - 1)^2 + (y - (-1))^2 &= 1. \end{aligned}$$

So the circle has center $(1, -1)$ and radius $\sqrt{1} = 1$.

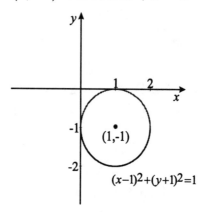

**Example 28** *(Exercise Set 1.3, Exercise 36) Find an equation of the circle with center $(1, 3)$ that passes through $(-2, 4)$.*

Solution: If $(-2, 4)$ is on the circle, then the line segment from the center point $(1, 3)$ to $(-2, 4)$ is a radius, and the length of the radius is

$$\begin{aligned} d((1,3),(-2,4)) &= \sqrt{(-2-1)^2 + (4-3)^2} \\ &= \sqrt{10}. \end{aligned}$$

The equation of the circle is

$$(x-1)^2 + (y-3)^2 = 10.$$

■

## 1.4 The Graph of an Equation

The graph of an equation is the set of all points $(x, y)$ that satisfy the equation. The variable $x$ is called the *independent* variable since its values can vary freely over a collection of real numbers. The variable $y$ is called the *dependent* variable since its value depends on the particular value of $x$ selected.

One way to obtain the graph of an equation is to plot several representative points that lie on the graph. The points where the graph crosses the $x$-axis and the $y$-axis, called the $x$- and $y$-intercepts, are particularly useful. This simple point-plotting approach will work in some cases but usually is not sufficient for our purposes in PreCalculus and Calculus.

## 1.4. THE GRAPH OF AN EQUATION

### 1.4.1 Plotting Points

**Example 29** *Sketch the graph of the given equation.*
(a) $y = x - 1$   (b) $y = x^2 + 1$

Solution:

(a) To sketch the graph we first make a table of representative points on the graph by selecting particular values of $x$ and then determining the associated $y$ values. The $x$-intercept is the point with $y$-coordinate zero. From the table we see that this is $(1, 0)$. The $y$-intercept is obtained by setting $x$ to 0 which gives the point $(0, -1)$. The figure shows that the graph of the equation is a straight line.

| $x$ | $y = x - 1$ |
|---|---|
| $-3$ | $-3 - 1 = -4$ |
| $-2$ | $-2 - 1 = -3$ |
| $-1$ | $-1 - 1 = -2$ |
| $0$ | $0 - 1 = -1$ |
| $1$ | $1 - 1 = 0$ |
| $2$ | $2 - 1 = 1$ |
| $3$ | $3 - 1 = 2$ |

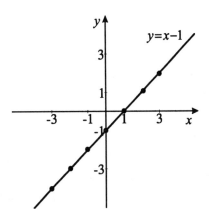

(b) Since $x^2 + 1 \geq 1$, the graph does not cross the $x$-axis and so it has no $x$-intercepts. The table shows that the $y$-intercept is $(0, 1)$. The graph is the parabola shown in the figure.

| $x$ | $y = x^2 + 1$ |
|---|---|
| $-3$ | $(-3)^2 + 1 = 10$ |
| $-2$ | $(-2)^2 + 1 = 5$ |
| $-1$ | $(-1)^2 + 1 = 2$ |
| $0$ | $0^2 + 1 = 1$ |
| $1$ | $1^2 + 1 = 2$ |
| $2$ | $2^2 + 1 = 5$ |
| $3$ | $3^2 + 1 = 10$ |

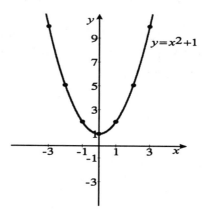

■

### 1.4.2 Symmetry

The left portion of the graph in part (b) of the proceeding example is the reflection through the $y$-axis of the right portion of the graph. *Symmeries* of this type are often useful in sketching graphs.

<u>Three types of symmetry</u>

1. A curve is *symmetric with respect to the y-axis* if its equation is unchanged when $x$ is replaced with $-x$. Geometrically this means that whenever the point $(x, y)$ is on the curve, the point $(-x, y)$ also is on the curve. For example, the graph of $y = x^2$.

2. A curve is *symmetric with respect to the x-axis* if its equation is unchanged when $y$ is replaced with $-y$. Geometrically this means that whenever the point $(x, y)$ is on the curve, the point $(x, -y)$ also is on the curve. For example, the graph of $x = y^2$.

## 1.4. THE GRAPH OF AN EQUATION

3. A curve is *symmetric with respect to the origin* if its equation remains unchanged when $x$ is replaced with $-x$ and $y$ is replaced with $-y$. Geometrically this means that whenever the point $(x, y)$ is on the curve, the point $(-x, -y)$ also is on the curve. For example, the graph of $y = x^3$.

**Example 30** *Test each curve for symmetry, find any $x$- and $y$-intercepts and sketch the graph of the equation.*
  (a) *(Exercise Set 1.4, Exercise 9)*  $y = x^2 - 1$
  (b) *(Exercise Set 1.4, Exercise 15)*  $y = x^3 + 1$

Solution:
(a)
y-axis symmetry: Yes, since if we replace $x$ with $-x$,

$$(-x)^2 - 1 = x^2 - 1$$

and the equation remains unchanged.

x-axis symmetry: No, since if $y$ is replaced with $-y$ the equation is changed. For example $(2, 3)$, is on the curve but $(2, -3)$ is not on the curve.

origin symmetry: No, since for example $(2, 3)$, is on the curve but $(-2, -3)$ is not on the curve.

x-intercepts: Solve the equation

$$\begin{aligned} x^2 - 1 &= 0 \\ x^2 &= 1 \\ x &= \pm 1. \end{aligned}$$

So the graph crosses the $x$-axis at the points $(-1, 0)$ and $(1, 0)$.

y-intercepts: Set $x = 0$, then $y = -1$ and the $y$-intercept is the point $(0, -1)$.

Since the graph is symmetric with respect to the $y$-axis we can plot the graph for positive values of $x$ and then reflect the graph about the $y$-axis as shown in the figure.

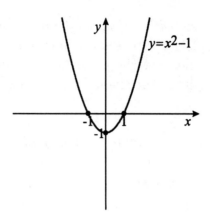

(b) If $x$ is replaced with $-x$, the equation becomes
$$y = (-x)^3 + 1 = -x^3 + 1$$
which has altered the original equation so the graph is not symmetric with respect to the $y$-axis. Replacing $y$ with $-y$ gives
$$\begin{aligned} -y &= x^3 + 1 \\ y &= -x^3 - 1, \end{aligned}$$
so the graph is not symmetric with respect to the $y$-axis. Replacing $x$ with $-x$ and $y$ with $-y$ gives
$$\begin{aligned} -y &= (-x)^3 + 1 = -x^3 + 1 \\ y &= x^3 - 1 \end{aligned}$$
and the curve is also not symmetric with respect to the origin. The curve is sketched in the figure. Notice that this particular curve is symmetric with respect to the point $(0, 1)$.

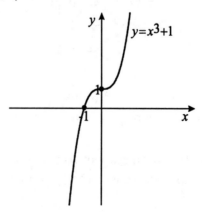

## 1.5. USING TECHNOLOGY TO GRAPH EQUATIONS

■

**Example 31** *(Exercise Set 1.4, Exercise 18) Sketch the graph of* $y = \dfrac{x^2 - x - 6}{x + 2}$.

Solution: First factor the numerator to get,

$$\begin{aligned} y &= \frac{x^2 - x - 6}{x + 2} \\ &= \frac{(x-3)(x+2)}{(x+2)} \\ &= x - 3, \text{ for } x \neq -2. \end{aligned}$$

It is important to realize that the original equation is not defined when $x$ is $-2$. This is illustrated graphically by removing the point $(-2, -5)$, from the graph of $y = x - 3$, as shown in the figure.

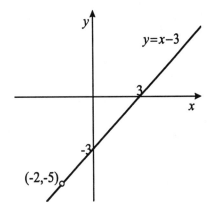

■

## 1.5 Using Technology to Graph Equations

Graphing devices sketch curves by plotting many points that are connected by very small line segments. The *viewing rectangle* of a graphing device is the rectangular portion of the plane in which the plot is displayed. Selecting the appropriate viewing rectangle when using a graphing device is very important. A viewing rectangle specified as $[a, b] \times [c, d]$, defines the rectangular region in the plane with $(x, y)$ restricted by $a \leq x \leq b$ and $c \leq y \leq d$.

**Example 32** *(Exercise Set 1.5, Exercise 3)* *Use a graphing device to sketch a graph of $y = x^3 - 20x + 25$ with the following viewing rectangles, and determine which gives the best representation for the graph of the equation.*

   *(a)* $[-2, 2] \times [-2, 2]$   *(b)* $[-5, 5] \times [-5, 5]$
   *(c)* $[-10, 10] \times [-70, 70]$   *(d)* $[-100, 100] \times [-200, 200]$

Solution: The graph in (a) appears to be a straight line segment, which we immediately reject as the graph of a third degree equation.

We expect the graph to be a smooth curve, so we also reject the graph in part (b).

The viewing rectangle in part (c) gives a nice smooth curve which we accept with confidence as a good representative of the graph.

The viewing rectangle used in part (d) is too large in both the $x$ and $y$ directions and clearly has distorted the graph. It does illustrate, however, that the curve in part (c) shows the important features of the graph.

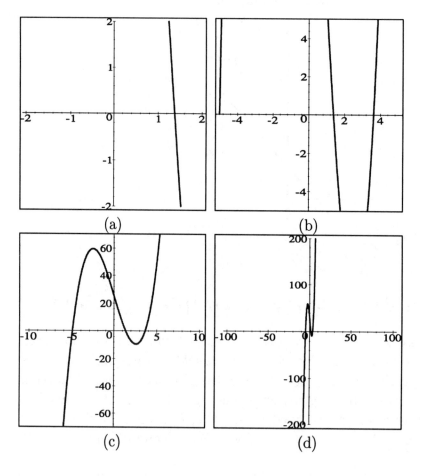

## 1.5. USING TECHNOLOGY TO GRAPH EQUATIONS

**Example 33** *Determine an appropriate viewing rectangle for the graph of the equation*
$$y = \frac{x+6}{x^2 - 1}.$$

Solution: We start by selecting a viewing rectangle of $[-5, 5] \times [-5, 5]$. The graph in part (a) appears to be a reasonable representation of the graph. However, we need to be careful when using a graphing device to be certain we have selected a viewing rectangle that shows all the important features of the graph. In this case, if we set $x = 0$, then the $y$-value of the point on the curve is $-6$, which we do not see on the graph in part (a).

In part (b) we have used a viewing rectangle of $[-5, 5] \times [-10, 5]$, and see we indeed missed a great deal of the curve in part (a).

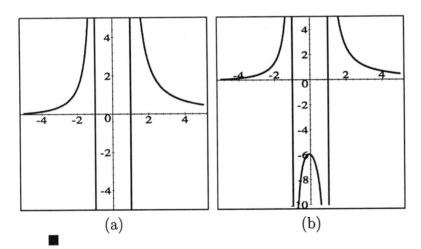

(a)  (b)

**Example 34** *(Exercise Set 1.5, Exercise 6(a)) Graphically approximate the solutions to the inequality $x^2 + 3x - 2 \geq 0$.*

Solution: In (a) the equation $y = x^2 + 3x - 2$ is graphed in a viewing rectangle $[-5, 5] \times [-5, 5]$. The inequality is positive, which corresponds to the portion of the graph that lies above the $x$-axis. Clicking on the $x$-intercepts of the graph gives values of $x \approx -3.63$ and $x \approx 0.56$. So, $x^2 + 3x - 2 \geq 0$ for approximately $x \leq -3.63$ or $x \geq 0.56$.

In (b) the viewing rectangle is $[0,1] \times [-1,1]$. Each tickmark between 0.5 and 0.6 is 0.04 apart, so our answer of 0.56 is accurate to within 2 decimal places. For more accuracy *zoom* closer in to the $x$-intercept by shrinking the viewing rectangle.

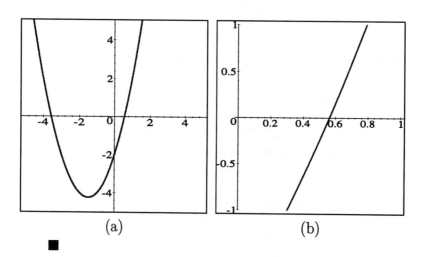

(a)         (b)

■

**Example 35** *(Exercise Set 1.5, Exercise 10) The number of bacteria in a culture at time $t$ is given by*

$$n = 10000 \left( \frac{3t^2 + 1}{t^2 + 1} \right).$$

*As the time $t$ increases, does the size of the bacteria colony become stable? If so, what is the stabilizing level?*

Solution: The equation along with the horizontal line at 30000 are plotted using a viewing rectangle of $[0, 25] \times [0, 32000]$. The figure indicates that the bacteria culture appears to level off to a stabilizing value of 30000.

## 1.5. USING TECHNOLOGY TO GRAPH EQUATIONS

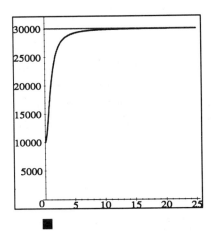

**Example 36** *(Exercise Set 1.5, Exercise 12)* Use a graphing device to plot a variety of curves of the form $y = (x-a)^2 + b$, where $a$ and $b$ are real numbers. Describe the effect that both positive and negative values of $a$ and $b$ have on the graph.

Solution: We separate the two parameters in figures (a) and (b). In (a) we plot $y = (x-a)^2$ for $a = -2, -1, 0, 1, 2$ and in (b) we plot $y = x^2 + b$ for $b = -2, -1, 0, 1, 2$. The parameter $a$ causes a horizontal shift in the graph of $y = x^2$, $a$ units to the right if $a > 0$ and $a$ units to the left if $a < 0$. The parameter $b$ causes a vertical shift in the graph of $y = x^2$, $b$ units up if $b > 0$ and $b$ units down if $b < 0$. The more general equation $y = (x-a)^2 + b$, involves both a horizontal and a vertical shift.

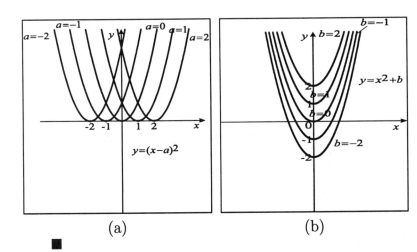

(a)       (b)

## 1.6 Functions

A *function* is a process that for each admissable input returns another *unique* real number. If the function is called $f$ and $x$ is an input value, then the output is denoted as $f(x)$, read "$f$ of $x$." Functions will play a central role in all quantitative study and you should become comfortable with the key ideas and the notation that is used.

### 1.6.1 Evaluation of Functions

**Example 37** *Find each value of the function $f$ defined by $f(x) = x^2 - 2x + 1$.*
  *(a) $f(3)$   (b) $f(-2)$   (c) $f\left(\frac{1}{2}\right)$*

Solution:
(a)
$$f(3) = (3)^2 - 2(3) + 1 = 4$$
(b)
$$f(-2) = (-2)^2 - 2(-2) + 1 = 9$$
(c)
$$f\left(\frac{1}{2}\right) = \left(\frac{1}{2}\right)^2 - 2\left(\frac{1}{2}\right) + 1$$
$$= \left(\frac{1}{4}\right) - 1 + 1$$
$$= \frac{1}{4}$$

■

The input to a function can be any expression that represents a real number. The variable used in the definition of the function is simply a place holder and evaluating a function at a specific value means replace the variable everywhere it appears with the input value. For example, if

$$f(x) = \frac{x^3 - x + 1}{2x - 1}$$

then the function can be thought of as

## 1.6. FUNCTIONS

$$f(\Box) = \frac{\Box^3 - \Box + 1}{2\Box - 1}.$$

Whatever fills the box on the left side of the equation must also fill the boxes on the right side.

**Example 38** (*Exercise Set 1.6, Exercise 2*) *If $f(t) = |t-2|$, find each of the following.*
  (a) $f(4)$   (b) $f(1)$   (c) $f(0)$   (d) $f(t+2)$   (e) $f(2-t^2)$   (f) $f(-t)$

Solution:
(a) $f(4) = |4-2| = 2$
(b) $f(1) = |1-2| = |-1| = 1$
(c) $f(0) = |0-2| = |-2| = 2$
(d) $f(t+2) = |(t+2) - 2| = |t|$
(e) $f(2-t^2) = |(2-t^2) - 2| = |-t^2| = t^2$
(f) $f(-t) = |-t-2| = |-(t+2)| = |t+2|$

■

### 1.6.2 Domain and Range

The *domain* of a function is the collection of real numbers that can be input to the function. The *range* is the collection of all outputs from the function. If the domain of a function is not explicitly stated we assume it is the largest set of real numbers for which the function is defined.

**Example 39** *Find the domain of the function.*
  (a) $f(x) = \dfrac{3}{x+1}$
  (b) (*Exercise Set 1.6, Exercise 17*) $f(x) = \sqrt{x(x-2)}$
  (c) $f(x) = \sqrt{\dfrac{1}{x^2 - 1}}$

Solution:

(a) The only values of $x$ for which the function is not defined are the values that make the denominator $x + 1 = 0$. So the domain is all real numbers except $x = -1$. The domain is $(-\infty, -1) \cup (-1, \infty)$.

(b) For the function to be defined, the expression under the radical must be greater than or equal to 0. To solve the inequality
$$x(x - 2) \geq 0$$
make the chart

| Interval | Test Value | Test | Inequality |
|---|---|---|---|
| $(-\infty, 0)$ | $-1$ | $-1(-1 - 2) = 3$ | $> 0$ |
| $(0, 2)$ | $1$ | $1(1 - 2) = -1$ | $< 0$ |
| $(2, \infty)$ | $3$ | $3(3 - 2) = 3$ | $> 0$ |

The table indicates that the inequality is positive on $(-\infty, 0) \cup (2, \infty)$. Since $\sqrt{0} = 0$, the values 0 and 2 are also in the domain, so the domain is $(-\infty, 0] \cup [2, \infty)$.

(c) This problem is similar to part (b), where the expression under the radical must be greater than or equal to zero but this time the values that make the denominator 0 are not included in the domain. So,
$$\begin{aligned} x^2 - 1 &= 0 \\ (x - 1)(x + 1) &= 0 \\ x &= \pm 1 \end{aligned}$$
and in the final answer 1 and $-1$ will not be in the domain. Now,
$$\begin{aligned} \frac{1}{x^2 - 1} &\geq 0 \\ x^2 - 1 &\geq 0 \\ (x - 1)(x + 1) &\geq 0 \end{aligned}$$
and

| Interval | Test Value | Test | Inequality |
|---|---|---|---|
| $(-\infty, -1)$ | $-2$ | $(-2 - 1)(-2 + 1) = 3$ | $> 0$ |
| $(-1, 1)$ | $0$ | $(0 - 1)(0 + 1) = -1$ | $< 0$ |
| $(1, \infty)$ | $2$ | $(2 - 1)(2 + 1) = 3$ | $> 0$ |

The domain is $(-\infty, -1) \cup (1, \infty)$.

■

## 1.6. FUNCTIONS

### 1.6.3 Vertical and Horizontal Line Tests

Suppose that you are given the graph of an equation. If every vertical line crosses the graph at most once, then the equation defines a function. If a vertical line crosses the graph of a function, then the $x$-intercept is in the domain. If a horizontal line crosses the graph of a function, then the $y$-intercept is in the range of the function.

**Example 40** *Determine if the graph defines a function and if so find the domain and the range.*

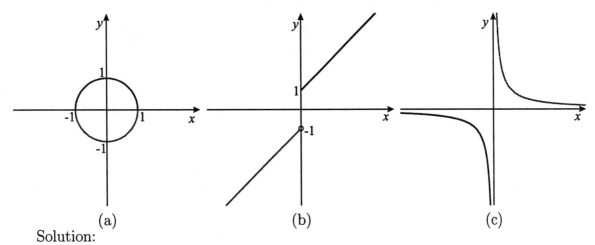

(a)  (b)  (c)

Solution:

(a) Any vertical line drawn between $x = -1$ and $x = 1$ crosses the curve in exactly two places, so the graph does not define a function.

In this example the curve is the circle with center the origin and radius 1, which has the equation $x^2 + y^2 = 1$. Solving for $y$ gives

$$y = \pm\sqrt{1 - x^2}$$

so

$$y = \sqrt{1 - x^2} \quad \text{or} \quad y = -\sqrt{1 - x^2}.$$

Each equation defines a function of $y$ in terms of $x$. Choosing the positive square root gives us a function that describes the upper semi-circle. Choosing the negative square root gives us a function that describes the lower semi-circle.

(b) Every vertical line crosses the curve in at most only one place, so the graph defines a function. Since every vertical line crosses the curve, the domain is the set of all real numbers, $(-\infty, \infty)$. The horizontal lines that do *not* cross the curve are those between $-1$ and $1$ including the line $y = -1$. The range is consequently $(-\infty, -1) \bigcup [1, \infty)$.

(c) The curve defines a function with domain all real numbers except $x = 0$. The range is also all real numbers except 0. So the domain is $(-\infty, 0) \bigcup (0, \infty)$ and the range is $(-\infty, 0) \bigcup (0, \infty)$.

∎

## 1.6.4 Difference Quotients

In calculus we frequently encounter an important quantity called the *difference quotient* of a function. The difference quotient for the function $f$ at a number $x$ in its domain is

$$\frac{f(x+h) - f(x)}{h},$$

where $h$ represents an arbitrary small positive number.

**Example 41** *(Exercise Set 1.6, Exercise 36) Let $f(x) = 4x^2 + 3x + 1$. Find*

$$f(x+h) \quad \text{and} \quad \frac{f(x+h) - f(x)}{h},$$

*assuming that $h \neq 0$.*

Solution: Since

$$\begin{aligned} f(x+h) &= 4(x+h)^2 + 3(x+h) + 1 \\ &= 4(x^2 + 2hx + h^2) + 3x + 3h + 1 \\ &= 4x^2 + 8hx + 4h^2 + 3x + 3h + 1, \end{aligned}$$

we have

$$\begin{aligned} \frac{f(x+h) - f(x)}{h} &= \frac{4(x+h)^2 + 3(x+h) + 1 - (4x^2 + 3x + 1)}{h} \\ &= \frac{4(x^2 + 2hx + h^2) + 3x + 3h + 1 - 4x^2 - 3x - 1}{h} \end{aligned}$$

## 1.6. FUNCTIONS

$$= \frac{4x^2 - 4x^2 + 3x - 3x + 1 - 1 + 8hx + 4h^2 + 3h}{h}$$

$$= \frac{8hx + 4h^2 + 3h}{h} = \frac{h(8x + 4h + 3)}{h}$$

$$= 8x + 4h + 3.$$

∎

**Example 42** Let $f(x) = \sqrt{2x+1}$. Find

$$f(x+h) \quad \text{and} \quad \frac{f(x+h) - f(x)}{h},$$

assuming that $h \neq 0$.

Solution: Since

$$f(x+h) = \sqrt{2x + 2h + 1},$$

we have

$$f(x+h) - f(x) = \sqrt{2x + 2h + 1} - \sqrt{2x+1}.$$

Multiplying the numerator and denominator by $\sqrt{2x+2h+1} + \sqrt{2x+1}$ permits us to rewrite the difference quotient as

$$\begin{aligned}
\frac{f(x+h) - f(x)}{h} &= \frac{\sqrt{2x+2h+1} - \sqrt{2x+1}}{h} \cdot \frac{\sqrt{2x+2h+1} + \sqrt{2x+1}}{\sqrt{2x+2h+1} + \sqrt{2x+1}} \\
&= \frac{\left(\sqrt{2x+2h+1}\right)^2 - \left(\sqrt{2x+1}\right)^2}{h\left(\sqrt{2x+2h+1} + \sqrt{2x+1}\right)} \\
&= \frac{2x + 2h + 1 - 2x - 1}{h\left(\sqrt{2x+2h+1} + \sqrt{2x+1}\right)} \\
&= \frac{2h}{h\left(\sqrt{2x+2h+1} + \sqrt{2x+1}\right)} \\
&= \frac{2}{\sqrt{2x+2h+1} + \sqrt{2x+1}}.
\end{aligned}$$

You might argue that this final result is not really simpler than the original form. In a sense this is true, but in calculus you will want to determine the value of the difference quotient as $h$ approaches zero. This final result shows that as $h$ approaches zero, the difference quotient approaches

$$\frac{2}{\sqrt{2x+0+1}+\sqrt{2x+1}} = \frac{2}{2\sqrt{2x+1}} = \frac{1}{\sqrt{2x+1}}.$$

■

### 1.6.5 Odd and Even Functions

A function $f$ is *odd* provided that for all $x$ in its domain $f(-x) = -f(x)$, and a function is *even* provided $f(-x) = f(x)$. Geometrically, this means that

> an odd function is symmetric with respect to the origin

and

> an even function is symmetric with respect to the $y$-axis.

**Example 43** *(Exercise Set 1.6, Exercise 43(a-d))* *Determine whether each of the following functions is even, odd, or neither even nor odd.*
(a) $f(x) = x^2 + 1$   (b) $f(x) = x^3 - 1$   (c) $f(x) = x^3 + 3x$   (d) $f(x) = \sqrt{x}$

Solution:
(a)

$$\begin{aligned} f(-x) &= (-x)^2 + 1 = x^2 + 1 = f(x), \text{ so } f \text{ is even.} \\ -f(x) &= -x^2 - 1 \neq f(-x), \text{ so } f \text{ is not odd.} \end{aligned}$$

(b)

$$\begin{aligned} f(-x) &= (-x)^3 - 1 = -x^3 - 1 \neq f(x), \text{ so } f \text{ is not even.} \\ -f(x) &= -x^3 + 1 \neq f(-x), \text{ so } f \text{ is not odd.} \end{aligned}$$

(c)

$$\begin{aligned} f(-x) &= (-x)^3 + 3(-x) = -x^3 - 3x \neq f(x), \text{ so } f \text{ is not even.} \\ -f(x) &= -x^3 - 3x = f(-x), \text{ so } f \text{ is odd.} \end{aligned}$$

## 1.6. FUNCTIONS

(d) Note that $f(x) = \sqrt{x}$ is defined for $x \geq 0$ and $f(-x) = \sqrt{-x}$ is defined for $x \leq 0$.

$$f(-x) = \sqrt{-x} \neq f(x), \text{ so } f \text{ is not even.}$$
$$-f(x) = -\sqrt{x} \neq f(-x), \text{ so } f \text{ is not odd.}$$

■

**Example 44** *(Exercise Set 1.6, Exercise 45)* *The graph of a function $f$ is given for $x \geq 0$. Extend the graph for $x < 0$ if*
(a) $f$ *is even*  (b) $f$ *is odd.*

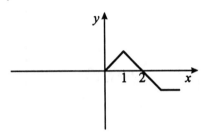

Solution:
(a) The graph must be symmetric with respect to the $y$-axis, which gives the following graph.

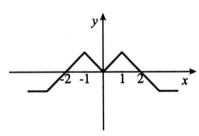

(b) The graph must be symmetric with respect to the origin, so the graph is as follows.

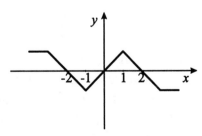

■

## 1.6.6 Applications

**Example 45** *(Exercise Set 1.6, Exercise 53)* *A rectangular plot of ground containing 432 ft² is to be fenced within a large plot.*

*(a) Express the perimeter of the plot as a function of the width. What is the domain of the function?*

*(b) Use a graphing device to approximate the dimensions of the plot that requires the least amount of fence.*

Solution:

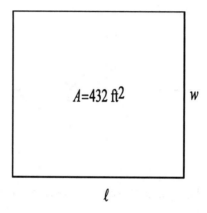

(a) The perimeter of the plot is
$$P = 2\ell + 2w.$$
To eliminate the variable $\ell$ from the equation, we use the information from the area of the plot. That is,
$$A = \ell w = 432$$
$$\ell = \frac{432}{w}.$$
Substituting into the equation for $P$ gives
$$P(w) = 2w + \frac{864}{w}.$$

(b) To find the exact dimensions of the plot that uses the least amount of fencing requires concepts from calculus. But we can approximate the dimensions using a graphing device to plot the perimeter $P$ with respect to the width $w$. The figure shows a low point at approximately $w \approx 20.6$ ft. That is, the perimeter is smallest, and hence the amount of fencing required is minimized, at this point. Then $\ell \approx \frac{432}{20.6} = 20.9$ ft.

## 1.6. FUNCTIONS

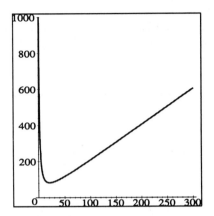

∎

**Example 46** *(Exercise Set 1.6, Exercise 54)* A manufacturer estimates the profit on producing $x$ units of their product at $P(x) = 300x - 2x^2$.

(a) What is the average rate of change in the profit as the number of units changes from $x$ to $x + h$?

(b) Use the result in part (a) to find the average rate of change in the profit as the number of units produced changes from 25 to 50?

(c) Sketch the graph of $y = P(x)$ and the line that passes through the points $(25, P(25))$ and $(50, P(50))$.

Solution:

(a) The average rate of change is the change in the values of the function as the independent variable varies over some interval. That is, the average rate of change is the difference quotient,

$$\begin{aligned}
\frac{P(x+h) - P(x)}{h} &= \frac{300(x+h) - 2(x+h)^2 - (300x - 2x^2)}{h} \\
&= \frac{300x + 300h - 2(x^2 + 2hx + h^2) - 300x + 2x^2}{h} \\
&= \frac{300h - 4hx - 2h^2}{h} = \frac{h(300 - 4x - 2h)}{h} \\
&= 300 - 4x - 2h.
\end{aligned}$$

(b) If $x = 25$ and $h = 25$, then $x + h = 50$, so the average rate of change can be computed using the final formula from part (a) as

$$300 - 4(25) - 2(25) = 150.$$

So, if the number of units produced increases from 25 to 50, the profit increases at an average rate of 150 per unit increase in production.

(c)

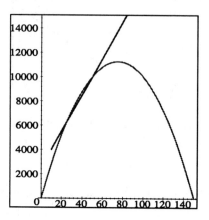

■

## 1.7 Linear Functions

### 1.7.1 The Slope of a Line

The *slope* of a line describes the inclination of the line and is a number that can be determined from any two points on the line. If two points on the line are $(x_1, y_1)$ and $(x_2, y_2)$, then the slope is

$$m = \frac{y_2 - y_1}{x_2 - x_1}.$$

It does not matter which point is considered first and which is second in the slope formula provided we are consistent in the order of subtraction in both the numerator and denominator. This follows from the fact that

$$\frac{y_1 - y_2}{x_1 - x_2} = \frac{-(y_2 - y_1)}{-(x_2 - x_1)} = \frac{y_2 - y_1}{x_2 - x_1}.$$

**Example 47** *Find the slope of the line that passes through the points and sketch the graph of the line.*

(a) $(5, 2)$ and $(-3, 1)$   (b) $(3, 4)$ and $(6, 1)$.

## 1.7. LINEAR FUNCTIONS

Solution:
(a) $m = \dfrac{1-2}{-3-5} = \dfrac{-1}{-8} = \dfrac{1}{8}$

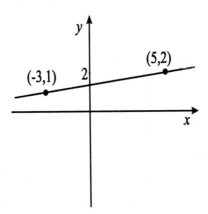

(b) $m = \dfrac{4-1}{3-6} = \dfrac{3}{-3} = -1$

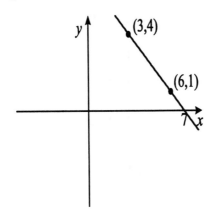

■

**Example 48** *Find the slope of the line that passes through the points and sketch the line.*
  (a) $(-2, -1)$ and $(7, -1)$   (b) $(-1, -4)$ and $(-1, 8)$.

Solution:
(a) $m = \dfrac{-1 - (-1)}{-2 - 7} = \dfrac{-1+1}{-9} = 0$

Since the slope of the line is 0 the line is horizontal. This line has $y$-intercept $(0, -1)$.

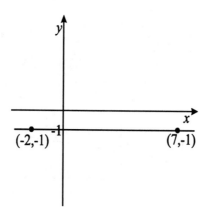

(b) $m = \dfrac{8+4}{-1+1} = \dfrac{12}{0}$ which is an undefined quantity.

Since the slope of the line is undefined the line is vertical. This line has $x$-intercept $(-1, 0)$.

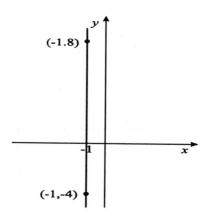

■

## 1.7.2 Point-Slope Equation of a Line

If a line passes through the point $(x_0, y_0)$ and has slope $m$, then a point $(x, y)$ will lie on the line precisely when

$$m = \frac{y - y_0}{x - x_0},$$

that is, when

$$y - y_0 = m(x - x_0).$$

## 1.7. LINEAR FUNCTIONS

This gives an equation describing all points that are on the line, called the *point-slope equation* of the line.

**Example 49** *(Exercise Set 1.7, Exercise 9) Find an equation of the line that passes through the point $(1, -2)$ and has slope 3.*

Solution: Substituting directly into the point-slope form gives

$$y - (-2) = 3(x - 1)$$
$$y = 3x - 5.$$

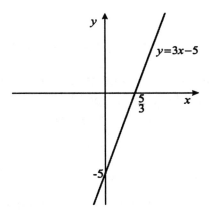

■

**Example 50** *Find the equation of the line passing through the two points $(-1, 1)$ and $(2, -3)$.*

Solution: First use the two points to find the slope. Then either of the given points along with the slope can be substituted into the point-slope form to find the equation. So

$$m = \frac{-3 - 1}{2 - (-1)} = -\frac{4}{3}$$
$$y - 1 = -\frac{4}{3}(x + 1)$$
$$y = -\frac{4}{3}x - \frac{1}{3}.$$

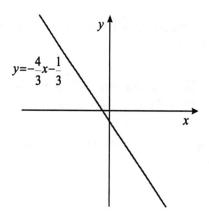

■

### 1.7.3 Slope-Intercept Equation of a Line

The *slope-intercept* equation of a line is a special case of the point-slope equation of the line where the $y$-intercept $(0, b)$ is the given point on the line. The slope-intercept equation then has the form

$$y = mx + b.$$

This was the form given in the final reduction in Example 49.

**Example 51** *(Exercise Set 1.7, Exercise 11) Find an equation of the line that has slope $-1$ and $y$-intercept 2.*

Solution: Substituting directly into the slope-intercept form of a line gives

$$y = -x + 2.$$

■

## 1.7. LINEAR FUNCTIONS

### 1.7.4 Parallel and Perpendicular Lines

Two lines are *parallel* provided they have the same slope, or both have undefined slopes (so are both vertical lines). Two lines are *perpendicular* if their slopes are negative reciprocals. That is, if the slopes are $m_1$ and $m_2$, then

$$m_1 m_2 = -1 \quad \text{or} \quad m_2 = -\frac{1}{m_1}.$$

**Example 52** *Find the line that passes through $(1,1)$ and is parallel to $2x + 3y = -2$. Sketch the two lines.*

Solution: First find the slope of the given line by writing the equation in the form $y = mx + b$. So,

$$\begin{aligned} 2x + 3y &= -2 \\ 3y &= -2x - 2 \\ y &= -\frac{2}{3}x - \frac{2}{3}, \end{aligned}$$

and the slope of the line is $-\frac{2}{3}$. Since the desired line is parallel to the given line, it also has slope $-\frac{2}{3}$. Since it passes through the point $(1,1)$ the equation is

$$\begin{aligned} y - 1 &= -\frac{2}{3}(x - 1) \\ y &= -\frac{2}{3}x + \frac{5}{3}. \end{aligned}$$

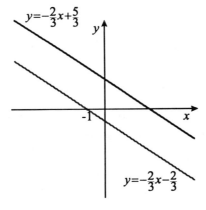

■

**Example 53** *Find the line that passes through $(-1, -1)$ and is perpendicular to $-3x + 5y = 3$. Sketch the two lines.*

Solution:
$$-3x + 5y = 3$$
$$5y = 3x + 3$$
$$y = \frac{3}{5}x + \frac{3}{5}.$$

The two lines are perpendicular, so the slope of the desired line is $-\frac{5}{3}$. The line we want passes through $(-1, -1)$ so its equation is

$$y + 1 = -\frac{5}{3}(x + 1)$$
$$y = -\frac{5}{3}x - \frac{8}{3}.$$

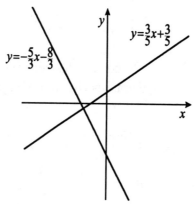

■

**Example 54** *(Exercise Set 1.7, Exercise 17) Find an equation of the line that is tangent to the circle $x^2 + y^2 = 3$ at the point $(1, \sqrt{2})$. At what other point on the circle will the tangent line be parallel to this line?*

Solution: The slope of the radius is $m = \frac{\sqrt{2}}{1} = \sqrt{2}$, so the tangent line has slope $-\frac{1}{\sqrt{2}} = -\frac{\sqrt{2}}{2}$. The equation of the tangent line is consequently

$$y - \sqrt{2} = -\frac{\sqrt{2}}{2}(x - 1)$$
$$y = -\frac{\sqrt{2}}{2}x + \frac{3\sqrt{2}}{2}.$$

## 1.7. LINEAR FUNCTIONS

The other point on the circle where the tangent line is parallel is at the point diametrically opposite $(1, \sqrt{2})$, which is the point $(-1, -\sqrt{2})$.

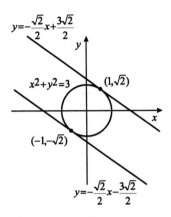

■

### 1.7.5 Applications

**Example 55** *(Exercise Set 1.7, Exercise 21) The average weight $W$, in grams, of a fish in a particular pond depends on the total number of fish in the pond according to the model*

$$W(n) = 500 - 0.5n$$

*(a) Sketch the graph of the function $W$.*
*(b) Express the total fish weight production in grams as a function of the number of fish in the pond.*
*(c) What happens when $n \geq 1000$?*

Solution:
(a) The easiest way to sketch the line is to find the intercepts.
<u>$y$-intercept:</u> Set $n = 0$, so $y = 500$.
<u>$x$-intercept:</u> Solve $500 - 0.5n = 0$, so $n = 1000$.

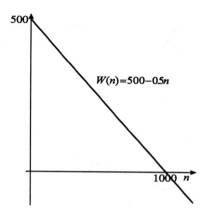

(b) The total fish production is given by

$$\begin{aligned} T(n) &= n \cdot W(n) \\ &= n(500 - 0.5n) \\ &= 500n - 0.5n^2 \end{aligned}$$

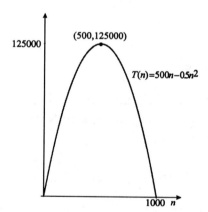

The figure indicates that the maximum sustainable fish population is $n = 500$.

(c) When $n = 1000$ the graph of the total weight crosses the $x$-axis so there are no longer any fish in the pond. This is the limiting value for the number of fish in the pond. For $n > 1000$, the graph will go below the $x$-axis and has no interpretation in the problem.
∎

**Example 56** *(Exercise Set 1.7, Exercise 22) A new computer workstation costs $ 10,000.00. Its useful lifetime is 5 years, at which time it will be worth* an estimated $ 2,000.00. The company calculates its depreciation using the linear decline method that is an option in the tax laws.

(a) Find a linear equation that expresses the value $V$ of the equipment as a function of time $t$, where $0 \leq t \leq 5$.

(b) How much will the equipment be worth after 2.5 years?

(c) What is the average rate of change in the value of the equipment from 1 to 3 years?

Solution:

(a) At time 0 the workstation is worth 10000 dollars and at time 5 it is worth 2000 dollars. If a linear model is used to represent value against time, the two points $(0, 10000)$ and $(5, 2000)$ lie on the line. The slope is

$$m = \frac{10000 - 2000}{0 - 5} = -\frac{8000}{5} = -1600,$$

and the equation of the line is

$$y = -1600t + 10000.$$

So the value at time $t$ is

$$V(t) = -1600t + 10000.$$

(b) Substitute $t = 2.5$ into the equation obtained in part (a) to get the value of the equipment after 2.5 years to be

$$V(2.5) = -1600(2.5) + 10000 = 6000 \text{ dollars}.$$

(c) The average rate of change is

$$\frac{V(3) - V(1)}{3 - 1} = \frac{5200 - 8400}{2} = -1600.$$

∎

## 1.8 Quadratic Functions

A *quadratic function* is a function of the form $f(x) = ax^2 + bx + c$, where $a, b,$ and $c$ are real numbers and $a \neq 0$. Every quadratic function can be rewritten in the *standard form* $f(x) = a(x-h)^2 + k$. The graph of a quadratic function is called a *parabola*.

## 1.8.1 Completing the Square

The process of writing a quadratic function in standard form is called *completing the square*.

**Example 57** *Write the quadratic function in standard form.*
(a) $f(x) = x^2 - 4x + 5$  (b) $f(x) = 2x^2 - 4x - 1$

Solution:
(a) To complete the square, take half the coefficient of the $x$ term, square it and both add and subtract the value, so the net result is to add 0. So,

$$\begin{aligned} f(x) &= x^2 - 4x + 5 \\ &= x^2 - 4x + \left(\frac{4}{2}\right)^2 - \left(\frac{4}{2}\right)^2 + 5 \\ &= x^2 - 4x + 4 + (-4 + 5) \\ &= (x-2)^2 + 1. \end{aligned}$$

(b) If the coefficient of the $x^2$ term is not 1, then first factor the coefficient from both the $x^2$ term and the $x$ term and proceed as in part (a). So,

$$\begin{aligned} f(x) &= 2x^2 - 4x - 1 \\ &= 2(x^2 - 2x) - 1 \\ &= 2\left(x^2 - 2x + \left(\frac{2}{2}\right)^2 - \left(\frac{2}{2}\right)^2\right) - 1 \\ &= 2\left(x^2 - 2x + 1 - 1\right) - 1 \\ &= 2(x-1)^2 - 2 - 1 \\ &= 2(x-1)^2 - 3. \end{aligned}$$

■

## 1.8.2 Horizontal and Vertical Shifts

If $h, k > 0$, then

the graph of $y = f(x - h)$ is the graph of $y = f(x)$ shifted $h$ units to the right,
the graph of $y = f(x + h)$ is the graph of $y = f(x)$ shifted $h$ units to the left,
the graph of $y = f(x) + k$ is the graph of $y = f(x)$ shifted $k$ units upward,
the graph of $y = f(x) - k$ is the graph of $y = f(x)$ shifted $k$ units downward.

## 1.8. QUADRATIC FUNCTIONS

**Example 58** *Use the graph of $y = f(x) = x^2$ given in the figure to sketch the graphs of*
(a) $y = f(x-1)$   (b) $y = f(x+2)$   (c) $y = f(x) + 1$
(d) $y = f(x) - 2$   (e) $y = f(x-2) + 3$.

Solution:

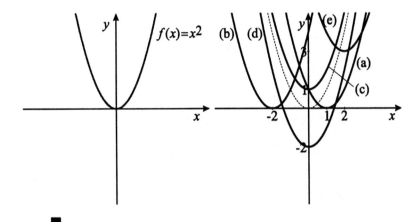

∎

**Example 59** *Sketch the graph of the quadratic function.*
(a) $f(x) = x^2 - 2x - 1$  (b) $f(x) = x^2 + 4x + 5$

Solution: To use the basic graph $y = x^2$ and the shifting properties, the quadratic function is first put in standard form by completing the square.
(a)

$$\begin{aligned} f(x) &= x^2 - 2x - 1 \\ &= (x^2 - 2x + 1 - 1) - 1 \\ &= (x-1)^2 - 2 \end{aligned}$$

To obtain the graph first shift $y = x^2$, to the right 1 unit and then 2 units downward. The vertex of the parabola is at $(1, -2)$.

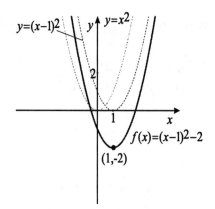

(b)
$$\begin{aligned} f(x) &= x^2 + 4x + 5 \\ &= (x^2 + 4x + 4 - 4) + 5 \\ &= (x+2)^2 + 1 \end{aligned}$$

To obtain the graph first shift $y = x^2$, to the left 2 units and then 1 unit upward. The vertex of the parabola is at $(-2, 1)$.

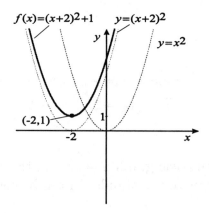

■

## 1.8.3 Horizontal and Vertical Scaling and Reflection

If $a > 1$, the graph of

$y = af(x)$, is a vertical stretching, by a factor of $a$, of the graph of $y = f(x)$,

$y = \dfrac{1}{a}f(x)$, is a vertical shrinking, by a factor of $a$, of the graph of $y = f(x)$.

## 1.8. QUADRATIC FUNCTIONS

The graph of $y = -f(x)$ is the reflection through the $x$-axis of the graph of $y = f(x)$.

For example, if $y = x^2$, then the point $(1, 1)$ on the curve lies directly below the point $(1, 2)$ on $y = 2x^2$,
directly above the point $(1, \frac{1}{2})$ on $y = \frac{1}{2}x^2$,
1 unit below the $x$-axis on $y = -x^2$.

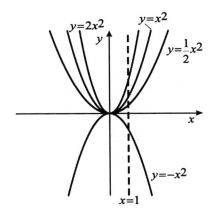

**Example 60** *Sketch the graph of the function and find the range.*
(a) (Exercise Set 1.8, Exercise 13)   $f(x) = 2x^2 - 8x + 10$
(b) (Exercise Set 1.8, Exercise 14)   $f(x) = -4x^2 - 4x + 3$

Solution:
(a) First we write

$$\begin{aligned} f(x) &= 2x^2 - 8x + 10 \\ &= 2(x^2 - 4x) + 10 \\ &= 2(x^2 - 4x + 4 - 4) + 10 \\ &= 2(x - 2)^2 + 2 \end{aligned}$$

To obtain the graph start with $y = x^2$, stretch it by a factor of 2, shift the result 2 units to the right and then 2 units upward. The range of the function is $[2, \infty)$ with a minimum point at $(2, 2)$.

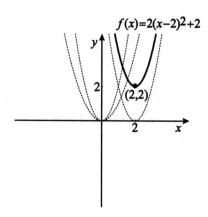

(b) Similarly,

$$\begin{aligned} f(x) &= -4x^2 - 4x + 3 \\ &= -4(x^2 + x) + 3 \\ &= -4\left(x^2 + x + \frac{1}{4} - \frac{1}{4}\right) + 3 \\ &= -4\left(x + \frac{1}{2}\right)^2 + 4 \end{aligned}$$

To obtain the graph start with $y = x^2$, stretch it by a factor of 4, reflect the result about the $x$-axis, shift the new result $\frac{1}{2}$ unit to the left, and then 4 units upward. The range of the function is $(-\infty, 4]$, with a maximum point at $\left(-\frac{1}{2}, 4\right)$.

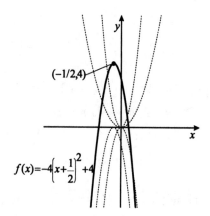

■

## 1.8. QUADRATIC FUNCTIONS

**Example 61** *(Exercise Set 1.8, Exercise 23) What can you say about $a, b$, and $c$ in $f(x) = ax^2 + bx + c$ if*
  *(a) $(1,1)$ is on the graph.*
  *(b) the y-intercept is 6.*
  *(c) $(1,1)$ is the vertex.*
  *(d) conditions (a), (b), and (c) are all satisfied.*

Solution:
(a) If $(1,1)$ is on the graph, then

$$\begin{aligned} 1 &= f(1) \\ 1 &= a(1)^2 + b(1) + c \\ a + b + c &= 1. \end{aligned}$$

(b) If $(0,6)$ is on the graph, then

$$\begin{aligned} 6 &= f(0) \\ 6 &= a(0)^2 + b(0) + c \\ c &= 6. \end{aligned}$$

(c) The vertex of a parabola is given by the point

$$\left( -\frac{b}{2a}, \frac{4ac - b^2}{4a} \right).$$

So

$$-\frac{b}{2a} = 1 \quad \text{and} \quad \frac{4ac - b^2}{4a} = 1.$$

(d) Using (a), (b), and (c),

$$\begin{aligned} a + b + c &= 1 \\ c &= 6 \\ b &= -2a \\ a - 2a + 6 &= 1 \\ -a &= -5 \\ a &= 5. \end{aligned}$$

So
$$a = 5$$
$$b = -10$$
$$c = 6$$
$$f(x) = 5x^2 - 10x + 6.$$

∎

### 1.8.4 Applications

**Example 62** *(Exercise Set 1.8, Exercise 29) National health care spending, in billions of dollars, has taken the shape of a parabola over the last few decades, increasing at an alarming rate. (See the table.)*

*(a) Using the 1965, 1980, and 1990 data only, fit a parabola of the form $y = a(x - 1965)^2 + b(x - 1965) + c$ to the data.*

*(b) What does the parabola predict the spreading will be in 2000?*

| Year | 1965 | 1970 | 1975 | 1980 | 1985 | 1990 |
|---|---|---|---|---|---|---|
| Dollars (in billions) | 30 | 80 | 120 | 250 | 400 | 690 |

Solution:

(a) Let $f(x) = a(x - 1965)^2 + b(x - 1965) + c$. Using the 1965, 1980 and 1990 data we have

$$30 = f(1965) = c$$
$$250 = f(1980) = 225a + 15b + c$$
$$690 = f(1990) = 625a + 25b + c$$

so

$$225a + 15b = 220$$
$$625a + 25b = 660.$$

To solve this system of two equations in the unknowns $a$ and $b$, begin by multiplying the first equation by 5 and the second by 3. Then subtract the second equation from the first. So

$$1125a + 75b = 1100$$

## 1.8. QUADRATIC FUNCTIONS

$$1875a + 75b = 1980$$
$$-750a = -880$$
$$a = \frac{-880}{-750}$$
$$a = \frac{88}{75}.$$

To find $b$, substitute the value $a = \frac{88}{75}$ into the first equation to get

$$225\left(\frac{88}{75}\right) + 15b = 220$$
$$3(88) + 15b = 220$$
$$15b = -44$$
$$b = -\frac{44}{15}.$$

So the parabola that fits the data is

$$f(x) = \frac{88}{75}(x-1965)^2 - \frac{44}{15}(x-1965) + 30.$$

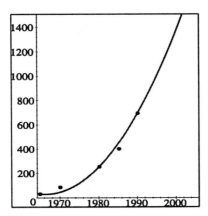

(b) In 2000 the equation predicts that the spreading will be approximately

$$f(2000) = \frac{4094}{3} \approx 1365 \text{ billion dollars.}$$

■

**Example 63** *(Exercise Set 1.8, Exercise 30) The profit function of a manufacturer when $x$ units of a commodity are produced and sold is given by*

$$P(x) = 200x - x^2$$

*(a) Sketch the graph of the profit function.*
*(b) How many units should be sold to yield a maximum profit?*
*(c) Compute the difference quotient $\frac{P(x+h)-P(x)}{h}$. What does the difference quotient approach as $h$ approaches 0? In economics this quantity is called the marginal profit when $x$ units are sold. It approximates the change in the profit when one additional unit is produced.*

Solution:
(a) First we write

$$\begin{aligned} P(x) &= -x^2 + 200x \\ &= -(x^2 - 200x) \\ &= -(x^2 - 200x + 10000 - 10000) \\ &= -(x - 100)^2 + 10000 \end{aligned}$$

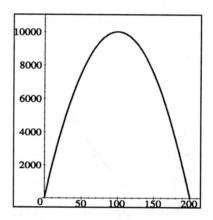

(b) The maximum profit is given by the $x$-coordinate of the highest point on the parabola in part (a), so is 100 units, with a maximum profit of $10,000.00.

(c)

$$\frac{P(x+h) - P(x)}{h} = \frac{200(x+h) - (x+h)^2 - (200x - x^2)}{h}$$

## 1.9. OTHER COMMON FUNCTIONS

$$= \frac{200x + 200h - x^2 - 2hx - h^2 - 200x + x^2}{h}$$

$$= \frac{200h - 2hx - h^2}{h} = \frac{h(200 - 2x - h)}{h}$$

$$= 200 - 2x - h$$

As $h$ approaches 0, the difference quotient approaches $200 - 2x$.

∎

## 1.9 Other Common Functions

### 1.9.1 The Absolute Value Function

The absolute value function is defined as

$$f(x) = |x| = \begin{cases} x, & \text{if } x \geq 0 \\ -x, & \text{if } x < 0. \end{cases}$$

For $x \geq 0$ the graph of the absolute value function is the same as the straight line $y = x$. For $x < 0$ the graph is the same as $y = -x$, as shown in the figure.

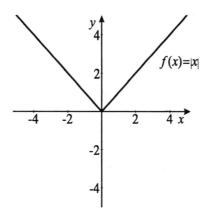

**Example 64** *Use the graph of $y = |x|$ to sketch the graph of the function.*
  *(a) (Exercise Set 1.9, Exercise 3) $f(x) = |x - 2| + 2$*
  *(b) (Exercise Set 1.9, Exercise 6) $f(x) = |2x - 5|$*

Solution:

(a) The shifting properties from the previous section applied to quadratic functions work equally well for any function. So the graph is just a horizontal shift of 2 units to the right of $y = |x|$ followed by a vertical shift 2 units upward.

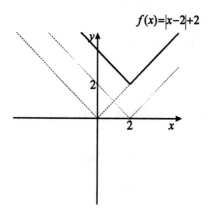

(b) First note that

$$\begin{aligned} f(x) &= |2x - 5| \\ &= 2\left|x - \frac{5}{2}\right|. \end{aligned}$$

The graph is a vertical stretching, by a factor of 2, of the graph of $y = |x|$, followed by a right shift $\frac{5}{2}$ units.

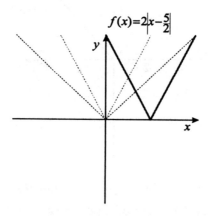

■

**Example 65** *Sketch the graph of $f(x) = |x^2 - 2x - 1|$.*

# 1.9. OTHER COMMON FUNCTIONS

Solution: First sketch the graph of $y = x^2 - 2x - 1$. Completing the square gives

$$\begin{aligned} y &= x^2 - 2x - 1 \\ &= x^2 - 2x + 1 - 1 - 1 \\ &= (x-1)^2 - 2. \end{aligned}$$

Taking the absolute value of this term will leave any portions of the graph that lie above the $x$-axis unchanged and will reflect about the $x$-axis any portions of the graph that lie below the $x$-axis.

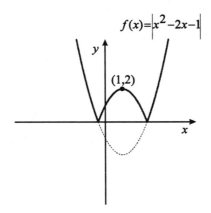

■

**Example 66** *(Exercise Set 1.9, Exercise 27) The Ohio Turnpike is 241 miles in length and has service plazas located 75 and 160 miles from Eastgate, the entrance to the Turnpike at the Pennsylvania line. Express the distance of a car from the nearest service plaza as a function of the car's distance from Eastgate, and sketch a graph of this function.*

Solution: Call the first plaza $A$ and the second plaza $B$. For the first 75 miles of the trip, the closest plaza is plaza $A$. The distance of the car from Eastgate varies from 0 miles to 75 miles and the distance to the plaza varies from 75 miles to 0 miles. If a linear model is used the line passes through the points, $(0, 75)$ and $(75, 0)$. The slope of the line is

$$m = \frac{75 - 0}{0 - 75} = -1$$

and the equation is

$$y = -x + 75 \quad \text{for} \quad 0 \leq x \leq 75,$$

where $x$ is the distance of the car from Eastgate and $y$ is the distance of the car from the closest plaza.

On the next 85 mile interval, the car is closest to plaza $A$ for the first 42.5 miles and closest to plaza $B$ for the second 42.5 miles. For the first 42.5 miles the distance of the car from Eastgate varies from 75 to 117.5 miles and its distance from plaza $A$ increases from 0 to 42.5 miles. So the linear model passes through the points $(75, 0)$ and $(117.5, 42.5)$. The slope of the line is

$$m = \frac{42.5 - 0}{117.5 - 75} = \frac{42.5}{42.5} = 1$$

and the equation is

$$y = x - 75 \quad \text{for } 75 \leq x \leq 117.5.$$

On the second half of the interval from 75 to 160, the line describing the position of the car passes through $(117.5, 42.5)$ and $(160, 0)$. The line has slope

$$m = -1$$

and equation

$$y = -x + 160.$$

On the final portion of the trip the car is between 160 and 241 miles from Eastgate and between 0 and $241 - 160 = 81$ miles from plaza $B$. The equation of the linear model is

$$y = x - 160.$$

The final function that expresses the distance of the car from the nearest plaza as a function of the car's distance from Eastgate is

$$f(x) = \begin{cases} -x + 75 & \text{if } 0 \leq x \leq 75 \\ x - 75 & \text{if } 75 \leq x \leq 117.5 \\ -x + 160 & \text{if } 117.5 \leq x \leq 160 \\ x - 160 & \text{if } 160 \leq x \leq 241 \end{cases} = \begin{cases} |x - 75| & \text{if } 0 \leq x \leq 117.5 \\ |x - 160| & \text{if } 117.5 \leq x \leq 241. \end{cases}$$

## 1.9. OTHER COMMON FUNCTIONS

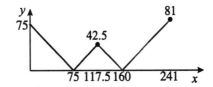

The graph indicates that we could also express the function as

$$f(x) = ||x - 117.5| - 42.5|.$$

∎

### 1.9.2  The Square Root Function

**Example 67** *(Exercise Set 1.9, Exercise 10) Use the graph of $y = \sqrt{x}$ to sketch the graph of the function $f(x) = 2 - \sqrt{x+2}$. Find the domain and range of the function.*

Solution: The graph of $y = -\sqrt{x}$ is the reflection of the graph of $y = \sqrt{x}$ about the $x$-axis . Then shift the graph of $y = -\sqrt{x}$, to the left 2 units and 2 units upward. The domain of the function is all real numbers such that

$$x + 2 \geq 0$$

that is,

$$x \geq -2.$$

In interval notation the domain is $[-2, \infty)$. From the graph, we see that the range is the interval $(-\infty, 2]$.

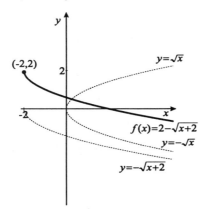

**Example 68** *(Exercise Set 1.9, Exercise 24)* Let $f(x) = x+1$ on the interval $[0, 4]$.

(a) Sketch the graph of $y = f(x)$ on the interval $[0, 4]$.

(b) Find an expression $d(x)$ for the distance from the origin to the point $(x, f(x))$. Use a graphing device to sketch the graph of $y = d(x)$ on the interval $[0, 4]$.

(c) Let $A(t)$ denote the area of the region bounded by the $x$-axis, the curve $y = f(x)$ and the vertical line $x = t$. Find an expression for $A(t)$ and sketch the graph of $y = A(t)$ on the interval $[0, 4]$.

Solution:
(a)

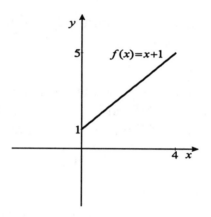

(b) The distance between the points $(0, 0)$ and $(x, f(x)) = (x, x+1)$ is
$$\begin{aligned} d(x) &= \sqrt{(x+1-0)^2 + (x-0)^2} \\ &= \sqrt{2x^2 + 2x + 1}. \end{aligned}$$

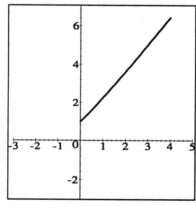

## 1.9. OTHER COMMON FUNCTIONS

(c) The region consists of a triangle sitting on top a rectangle. So

$$\begin{aligned} A(t) &= 1(t) + \frac{1}{2}t(t+1-1) \\ &= \frac{1}{2}t^2 + t. \end{aligned}$$

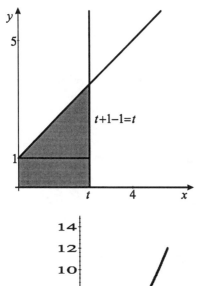

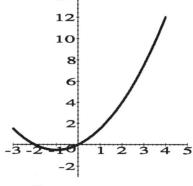

■

### 1.9.3 The Greatest Integer Function

The greatest integer function is defined by

$$\lfloor x \rfloor = \text{the largest integer less than or equal to } x.$$

**Example 69** *Evaluate each of the following.*
(a) $\lfloor 2 \rfloor$ (b) $\lfloor 1.6 \rfloor$ (c) $\lfloor -4 \rfloor$ (d) $\lfloor -3.5 \rfloor$

Solution:

(a) The greatest integer less than or equal to an integer is the integer itself, so $\lfloor 2 \rfloor = 2$.

(b) Since $1 \leq 1.6 < 2$, we have $\lfloor 1.6 \rfloor = 1$.

(c) $-4$ is an integer, so $\lfloor -4 \rfloor = -4$.

(d) Since $-4 \leq -3.5 < -3$, and the smaller integer is $-4$, we have $\lfloor -3.5 \rfloor = -4$.

∎

**Example 70** *Sketch the graph of* $f(x) = x + \lfloor x \rfloor$.

Solution: First write the definition for $f(x)$ on a selection of intervals. The meaning of $f(x)$ depends on the values of $\lfloor x \rfloor$. Since

$$\lfloor x \rfloor = \begin{cases} 0, & \text{if } 0 \leq x < 1, \\ 1, & \text{if } 1 \leq x < 2, \\ 2, & \text{if } 2 \leq x < 3, \\ 3, & \text{if } 3 \leq x < 4, \\ -1, & \text{if } -1 \leq x < 0, \\ -2, & \text{if } -2 \leq x < -1, \end{cases}$$

$$f(x) = x + \lfloor x \rfloor = \begin{cases} x, & \text{if } 0 \leq x < 1, \\ x+1, & \text{if } 1 \leq x < 2, \\ x+2, & \text{if } 2 \leq x < 3, \\ x+3, & \text{if } 3 \leq x < 4, \\ x-1, & \text{if } -1 \leq x < 0, \\ x-2, & \text{if } -2 \leq x < -1. \end{cases}$$

If $a$ represents an integer, then the function is equivalent to

$$f(x) = x + a, \text{ for } a \leq x < a+1.$$

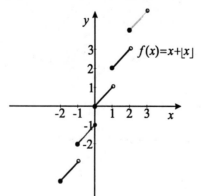

## 1.10 Arithmetic Combinations of Functions

If $f$ and $g$ are functions and $x$ is an admissable input for each of the functions, then $f(x)$ and $g(x)$ are both real numbers. The quantities $f(x)+g(x)$, $f(x)-g(x)$, $f(x) \cdot g(x)$, and if $g(x) \neq 0$, $\frac{f(x)}{g(x)}$, are all real numbers. We can define the *arithmetic combinations* of two functions by the formulas

$$
\begin{aligned}
\text{Addition} &: (f+g)(x) = f(x) + g(x) \\
\text{Subtraction} &: (f-g)(x) = f(x) - g(x) \\
\text{Multiplication} &: (fg)(x) = f(x) \cdot g(x) \\
\text{Division} &: \left(\frac{f}{g}\right)(x) = \frac{f(x)}{g(x)}
\end{aligned}
$$

As you must expect, the symbols $f+g, f-g, fg$, and $\frac{f}{g}$ are the names used for the new functions, and the expressions on the right of the equation indicate how to evaluate these functions at specific inputs.

For $x$ to be an admissable input value for $f+g$, $f-g$ or $fg$, the value $x$ must be a valid input for both $f$ and $g$. So if the domain of the function $f$ is $A$ and the domain of $g$ is $B$, the domain of $f+g$, $f-g$ and $fg$ consists of all real numbers common to both domains which is the intersection $A \cap B$. The domain of $\frac{f}{g}$ consists of those values in $A \cap B$, excluding any values of $x$ for which $g(x) = 0$.

**Example 71** *Let $f(x) = x^2$ and $g(x) = x+1$. Find $f+g, f-g, fg$, and $f/g$ and give the domains of each new function.*

Solution:

$$
\begin{aligned}
(f+g)(x) &= f(x) + g(x) = x^2 + x + 1 \\
(f-g)(x) &= f(x) - g(x) = x^2 - (x+1) = x^2 - x - 1 \\
(fg)(x) &= f(x) \cdot g(x) = x^2(x+1) = x^3 + x^2 \\
\left(\frac{f}{g}\right)(x) &= \frac{f(x)}{g(x)} = \frac{x^2}{x+1}
\end{aligned}
$$

Since the domain of $f$ is the set of all real numbers and the domain of $g$ is the set of all real numbers, the domain of $f+g, f-g$ and $fg$ is the set of all real numbers. The domain of $f/g$ is the set of all real numbers satisfying $x + 1 \neq 0$, that is, the domain is $(-\infty, -1) \cup (-1, \infty)$.

■

**Example 72** Let $f(x) = \frac{1}{x}$ and $g(x) = \sqrt{x+2}$. Find $f+g, f-g, fg$, and $f/g$ and give the domains of each new function.

Solution:

$$(f+g)(x) = f(x) + g(x) = \frac{1}{x} + \sqrt{x+2}$$

$$(f-g)(x) = f(x) - g(x) = \frac{1}{x} - \sqrt{x+2}$$

$$(fg)(x) = f(x) \cdot g(x) = \frac{1}{x} \cdot \sqrt{x+2} = \frac{\sqrt{x+2}}{x}$$

$$\left(\frac{f}{g}\right)(x) = \frac{f(x)}{g(x)} = \frac{\frac{1}{x}}{\sqrt{x+2}} = \frac{1}{x\sqrt{x+2}}$$

Domain of $f$: All real numbers except $x = 0$, that is, $(-\infty, 0) \cup (0, \infty)$.
Domain of $g$: All $x$ for which $\sqrt{x+2}$ is defined, so

$$x + 2 \geq 0.$$

That is,

$$x \geq -2.$$

Domain of $f+g, f-g, fg$: The intersection of the domains of $f$ and $g$, the set $[-2, 0) \cup (0, \infty)$.

Domain of $f/g$: All $x$ in $[-2, 0) \cup (0, \infty)$., except those values that make the denominator of $f/g$ equal to 0. Since

$$x\sqrt{x+2} = 0 \Rightarrow$$
$$x = 0, \quad x = -2,$$

the domain is $(-2, 0) \cup (0, \infty)$.

■

## 1.10. ARITHMETIC COMBINATIONS OF FUNCTIONS

**Example 73** Let $f(x) = x^2 + x - 2$ and $g(x) = x - 1$. Find $f+g, f-g, fg,$ and $f/g$ and give the domains of each new function.

Solution:

$$\begin{aligned}
(f+g)(x) &= f(x) + g(x) = x^2 + x - 2 + x - 1 = x^2 + 2x - 3 \\
(f-g)(x) &= f(x) - g(x) = x^2 + x - 2 - (x-1) \\
&= x^2 + x - 2 - x + 1 = x^2 - 1 \\
(fg)(x) &= f(x) \cdot g(x) = (x^2 + x - 2)(x - 1) \\
&= x^3 - x^2 + x^2 - x - 2x + 2 = x^3 - 3x + 2 \\
\left(\frac{f}{g}\right)(x) &= \frac{f(x)}{g(x)} = \frac{x^2 + x - 2}{x - 1} = \frac{(x-1)(x+2)}{x-1} = x + 2
\end{aligned}$$

<u>Domains of $f$ and $g$</u> : All real numbers.
<u>Domain of $f+g, f-g, fg$</u> : All real numbers.

<u>Domain of $f/g$</u> : It appears from the simplification for $f/g$, that the domain is all real numbers. This is *not* the case since for $x$ to be in the domain of $f/g$ it must be in both the domains of $f$ and $g$ and $g(x)$ *must be nonzero*. So $x = 1$, must be removed from the domain of $f/g$, since $g(1) = 0$. The domain of $f/g$ is $(-\infty, 1) \cup (1, \infty)$.
∎

**Example 74** Let
$$f(x) = \begin{cases} x^2 + 1 & \text{if } x \geq 0 \\ x & \text{if } x < 0 \end{cases}$$

and
$$g(x) = \begin{cases} x + 1 & \text{if } x \geq 0 \\ -1 & \text{if } x < 0 \end{cases}.$$

Find $f+g, f-g, fg,$ and $f/g$ and give the domains of each new function.

Solution: The functions are plotted in the figure.

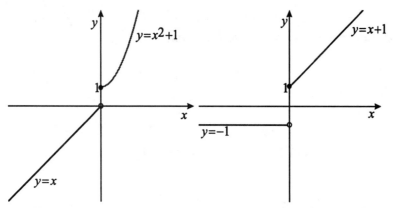

To combine these two functions, consider the definitions in two separate parts of the domains of $f$ and $g$, for $x \geq 0$ and for $x < 0$. So,

$$(f+g)(x) = \begin{cases} x^2 + x + 2 & \text{if } x \geq 0 \\ x - 1 & \text{if } x < 0 \end{cases}$$

$$(f-g)(x) = \begin{cases} x^2 - x & \text{if } x \geq 0 \\ x + 1 & \text{if } x < 0 \end{cases}$$

$$(fg)(x) = \begin{cases} x^3 + x^2 + x + 1 & \text{if } x \geq 0 \\ -x & \text{if } x < 0 \end{cases}$$

$$\left(\frac{f}{g}\right)(x) = \begin{cases} \frac{x^2+1}{x+1} & \text{if } x \geq 0 \\ -x & \text{if } x < 0. \end{cases}$$

Domains of $f$ and $g$: All real numbers, since all real numbers are specified in the two part definitions.

Domain of $f + g, f - g, fg$: All real numbers.

Domain of $f/g$: The quotient in the first part of the definition is undefined when $x = -1$. However, this part of the definition is only valid for $x \geq 0$, so this is not a problem. Notice this can also be seen from the fact the $g(x)$ is never zero. The domain is the set of all real numbers.

■

**Example 75** Let $f(x) = \sqrt{x^2 - 1}$ and $g(x) = \sqrt{9 - x^2}$. Find $f+g, f-g, fg$, and $f/g$ and give the domains of each new function.

Solution: First we have

$$(f+g)(x) = f(x) + g(x) = \sqrt{x^2 - 1} + \sqrt{9 - x^2}$$
$$(f-g)(x) = f(x) - g(x) = \sqrt{x^2 - 1} - \sqrt{9 - x^2}$$

## 1.10. ARITHMETIC COMBINATIONS OF FUNCTIONS

$$(fg)(x) = f(x) \cdot g(x) = \sqrt{x^2 - 1} \cdot \sqrt{9 - x^2}$$

$$\left(\frac{f}{g}\right)(x) = \frac{f(x)}{g(x)} = \frac{\sqrt{x^2 - 1}}{\sqrt{9 - x^2}}$$

Domain of $f$: The expression under the radical must be greater than or equal to 0. To solve the inequality factor and set up a table. So solve

$$x^2 - 1 = (x - 1)(x + 1) \geq 0.$$

| Interval | Test Value | Test | Inequality |
|---|---|---|---|
| $(-\infty, -1)$ | $-2$ | $(-2-1)(-2+1) = 3$ | $> 0$ |
| $(-1, 1)$ | $0$ | $(0-1)(0+1) = -1$ | $< 0$ |
| $(1, \infty)$ | $2$ | $(2-1)(2+1) = 3$ | $> 0$ |

Since $x^2 - 1 = 0$, for $x = \pm 1$, the domain of $f$ is $(-\infty, -1] \cup [1, \infty)$.

Domain of $g$: The expression under the radical must be greater than or equal to 0. So solve

$$9 - x^2 = (3 - x)(3 + x) \geq 0.$$

| Interval | Test Value | Test | Inequality |
|---|---|---|---|
| $(-\infty, -3)$ | $-4$ | $(3+4)(3-4) = -7$ | $< 0$ |
| $(-3, 3)$ | $0$ | $(3-0)(3+0) = 9$ | $> 0$ |
| $(3, \infty)$ | $4$ | $(3-4)(3+4) = -7$ | $< 0$ |

Since $9 - x^2 = 0$, for $x = \pm 3$, the domain of $g$ is $[-3, 3]$.

Domain of $f + g, f - g, fg$: $[-3, -1] \cup [1, 3]$.
Domain of $f/g$: Remove the values which make the denominator 0. So

$$\sqrt{9 - x^2} = 0$$
$$9 - x^2 = 0$$
$$x = \pm 3.$$

The domain of $f/g$ is $(-3, -1] \cup [1, 3)$.
■

**Example 76** *(Exercise Set 1.10, Exercise 11) Functions $f$ and $g$ are defined by $f(x) = x^2 - 4$ and $g(x) = x + 2$. Sketch the graph of $y = f(x)/g(x)$. Sketch the graph of $y = x + 2$. How do the graphs differ?*

Solution:

$$\frac{f(x)}{g(x)} = \frac{x^2 - 4}{x - 2}$$
$$= \frac{(x+2)(x-2)}{x-2} = x + 2, \text{ for } x \neq 2$$

The important observation here is that the quotient function is not defined at $x = 2$ since $g(2) = 0$. The fact that the fraction simplifies does not change the fact that $\frac{x^2-4}{x-2}$ is undefined for $x = 2$. The only difference in the graphs of $y = x + 2$ and $y = \frac{f(x)}{g(x)}$, is that the point $(2, 4)$ is removed from the graph of $y = \frac{f(x)}{g(x)}$.

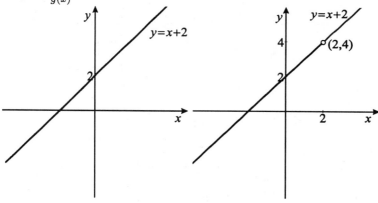

■

## 1.10.1 The Reciprocal Graphing Technique

Properties of the function $g(x) = \frac{1}{f(x)}$, called the *reciprocal of f*, can be obtained from the graph of the function $f$. The most important values of $x$ are those that make $f(x) = 0$.

**Example 77** *(Exercise Set 1.10, Exercise 12(d))* Let $g(x) = x^2 - 4x + 3$. *Use the results about the graph of the reciprocal of a function to sketch the graph of* $h(x) = 1/g(x)$.

Solution: The points where $g(x) = 0$, which is where the graph of $y = g(x)$ crosses the $x$-axis, are the points where $h(x)$ is undefined and are the essential points in determining the graph. So

$$x^2 - 4x + 3 = 0$$

## 1.10. ARITHMETIC COMBINATIONS OF FUNCTIONS

$$(x-3)(x-1) = 0$$
$$x = 3, x = 1.$$

The graph of $g(x) = x^2 - 4x + 3 = (x-2)^2 - 1$ in the figure shows that

$$g(x) > 0, \text{ for } x > 3 \text{ or } x < 1$$
$$g(x) < 0, \text{ for } 1 < x < 3.$$

As $x$ gets close to 1 or 3, $g(x)$ gets close to 0, so the magnitude of $h(x) = \frac{1}{g(x)}$ becomes arbitrarily large and

$$x \to 1^+, h(x) \to -\infty$$
$$x \to 1^-, h(x) \to \infty$$
$$x \to 3^+, h(x) \to \infty$$
$$x \to 3^-, h(x) \to -\infty.$$

As $g(x)$ gets large in magnitude, $h(x) = \frac{1}{g(x)}$ gets very small. So,

$$x \to \infty, g(x) \to \infty \quad \text{and} \quad h(x) \to 0$$
$$x \to -\infty, g(x) \to \infty \quad \text{and} \quad h(x) \to 0.$$

The vertex of the parabola is $(2, -1) = (2, g(2))$, so the point $\left(2, \frac{1}{g(2)}\right) = (2, -1)$ is also on the graph of $y = h(x)$. This is generally a good additional point to plot.

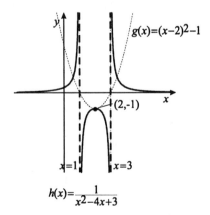

**Example 78** *(Exercise Set 1.10, Exercise 14)* Sketch the graphs of the given functions in the order given, and observe the difference in the graph that each successive complication introduces.

(a) $f_1(x) = x - 2$
(b) $f_2(x) = (x-2)^2$
(c) $f_3(x) = x^2 - 4x + 2$
(d) $f_4(x) = |x^2 - 4x + 2|$
(e) $f_5(x) = \dfrac{1}{|x^2 - 4x + 2|}$
(f) $f_5(x) = \dfrac{2}{|x^2 - 4x + 2|}$

Solution: (a) The graph of $y = f_1(x)$ is a straight line passing through $(2,0)$.

(b) The graph of $y = f_2(x)$ is a parabola with vertex at $(2,0)$.
(c) Since

$$\begin{aligned} f_3(x) &= x^2 - 4x + 2 \\ &= x^2 - 4x + 4 - 4 + 2 \\ &= (x-2)^2 - 2, \end{aligned}$$

the graph of $y = f_3(x)$ is obtained by shifting the graph of $y = f_2(x)$ downward 2 units.

(d) The graph of $y = f_4(x) = |x^2 - 4x + 2|$ is obtained by reflecting any portions of $y = f_3(x) = x^2 - 4x + 2$ that lie below the $x$-axis above.

(e) To obtain the graph of

$$y = f_5(x) = \frac{1}{|x^2 - 4x + 2|}$$

use the reciprocal graphing technique. The the graph of

$$y = f_6(x) = \frac{2}{|x^2 - 4x + 2|}$$

is a vertical scaling, by a factor of 2, of the graph of

$$y = f_5(x) = \frac{1}{|x^2 - 4x + 2|}.$$

## 1.11. COMPOSITION OF FUNCTIONS

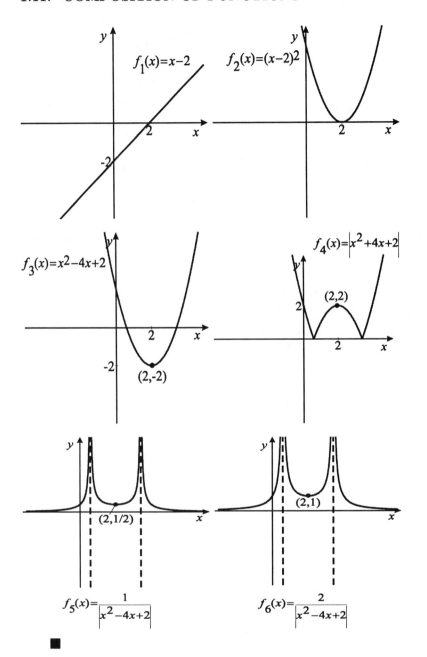

## 1.11 Composition of Functions

The *composition* of two functions is a special way of combining the two processes. For example, the function given by $f(x) = (x^2 + 1)^3$ can be interpreted as

$$x \to \boxed{\square^2+1} \to x^2+1 \to \boxed{[x^2+1]^3} \to (x^2+1)^3.$$

So $x$ is input to the process that squares the number and adds one, and then this new value is used as input to the process that cubes a number. The final result is the process that does both operations in the prescribed order.

The composition of the two functions $f$ and $g$ is written as

$$(f \circ g)(x) = f(g(x))$$

with the domain being all $x$ in the domain of $g$ so that $g(x)$ is in the domain of $f$.

**Example 79** Let $f(x) = x^2 + 1$ and $g(x) = \sqrt{x}$.
(a) Find $(f \circ g)(2)$ and $(g \circ f)(-3)$.
(b) Write expressions for $f(g(x))$ and $g(f(x))$.
(c) Can $f \circ g$ be evaluated at $-2$?

Solution:
(a) $\underline{(f \circ g)(2)}$ : First evaluate $g$ at 2, so $g(2) = \sqrt{2}$. Then substitute this value into $f$, so

$$\begin{aligned}(f \circ g)(2) &= f(g(2)) \\ &= f(\sqrt{2}) = \left(\sqrt{2}\right)^2 + 1 \\ &= 3.\end{aligned}$$

$\underline{(g \circ f)(-3)}$ :

$$\begin{aligned}(g \circ f)(-3) &= g(f(-3)) \\ &= g(10) \\ &= \sqrt{10}\end{aligned}$$

(b)

$$\begin{aligned}f(g(x)) &= f(\sqrt{x}) \\ &= \left(\sqrt{x}\right)^2 + 1 \\ &= x + 1\end{aligned}$$

## 1.11. COMPOSITION OF FUNCTIONS

$$g(f(x)) = g(x^2 + 1)$$
$$= \sqrt{x^2 + 1}$$

(c) No, $f \circ g$ can not be evaluated at $-2$, since $-2$ is not in the domain of $g$.

■

**Example 80** *Let $f(x) = \sqrt{x - 6}$ and $g(x) = x^2 + 5x$. Find $f(g(x))$ and specify the domain of $f \circ g$.*

Solution:

$$f(g(x)) = f(x^2 + 5x)$$
$$= \sqrt{x^2 + 5x - 6}$$

The domain of $f \circ g$ is the set of all $x$ in the domain of $g$, so that $g(x)$ is also in the domain of $f$. The domain of $g$ is all real numbers and the domain of $f$ is the interval $[6, \infty)$. So $x$ is in the domain of $f \circ g$ provided that

$$x^2 + 5x \geq 6$$
$$x^2 + 5x - 6 \geq 0$$
$$(x - 1)(x + 6) \geq 0$$

To solve the inequality we have,

| Interval | Test Value | Test | Inequality |
|---|---|---|---|
| $(-\infty, -6)$ | $-7$ | $(-7 - 1)(-7 + 6) = 8$ | $> 0$ |
| $(-6, 1)$ | $0$ | $(0 - 1)(0 + 6) = -6$ | $< 0$ |
| $(1, \infty)$ | $2$ | $(2 - 1)(2 + 6) = 8$ | $> 0$ |

The domain of $f \circ g$ is $(-\infty, -6] \cup [1, \infty)$.

■

**Example 81** *Write each function $h$ as a composition $f \circ g$.*
*(a) $h(x) = (x^5 + 3x^3 - 2x + 1)^8$  (b) $h(x) = \sqrt{x^2 - 2x + 1}$*

Solution:
(a) A natural separation of the process $h$ is to first perform the *inside* operation of $x^5 + 3x^3 - 2x + 1$ and then performing the *outside* operation of

raising a number to the 8th power. Since we want $g$ to be the inside operation and $f$ to be the outside operation we have,

$$\begin{aligned} f(x) &= x^8 \\ g(x) &= x^5 + 3x^3 - 2x + 1 \\ f(g(x)) &= (x^5 + 3x^3 - 2x + 1)^8 \\ &= h(x). \end{aligned}$$

(b) Let
$$f(x) = \sqrt{x} \quad \text{and} \quad g(x) = x^2 - 2x + 1$$

Then
$$\begin{aligned} f(g(x)) &= \sqrt{x^2 - 2x + 1} \\ &= h(x). \end{aligned}$$

■

**Example 82** *(Exercise Set 1.11, Exercise 18) Show that the composition of two odd functions is an odd function.*

Solution: A function $f$ is an odd function if $f(-x) = -f(x)$. If $f$ and $g$ are both odd functions, then

$$\begin{aligned} f(-x) &= -f(x) \\ g(-x) &= -g(x). \end{aligned}$$

To show the composition is an odd function it is necessary to verify that we have
$$(f \circ g)(x) = -(f \circ g)(x).$$

So
$$\begin{aligned} (f \circ g)(x) &= f(g(-x)) \\ &= f(-g(x)) \\ &= -f(g(x)) \\ &= -(f \circ g)(x). \end{aligned}$$

■

## 1.11. COMPOSITION OF FUNCTIONS

**Example 83** *(Exercise Set 1.11, Exercise 22) What type of function is the composition of (a) two linear functions? (b) two quadratic functions? (c) a linear and a quadratic function?*

Solution:
(a) Let $f(x) = ax + b$ and $g(x) = cx + d$. Then
$$\begin{aligned} f(g(x)) &= a(cx + d) + b \\ &= acx + (ad + b) \end{aligned}$$
which is a linear function. The graph of $f \circ g$ has slope $ac$.

(b) Let $f(x) = ax^2 + bx + c$ and $g(x) = \alpha x^2 + \beta x + \gamma$. Then
$$\begin{aligned} f(g(x)) &= a(\alpha x^2 + \beta x + \gamma)^2 + b(\alpha x^2 + \beta x + \gamma) + c \\ &= a\alpha^2 x^4 + 2a\alpha x^3 \beta + 2a\alpha x^2 \gamma + a\beta^2 x^2 + 2a\beta x \gamma \\ &\quad + a\gamma^2 + b\alpha x^2 + b\beta x + b\gamma + c \end{aligned}$$
which has highest power of $x$ equal to 4, that is, a *quartic* function.

(c) Let $f(x) = ax + b$ and $g(x) = \alpha x^2 + \beta x + \gamma$. Then
$$\begin{aligned} f(g(x)) &= a(\alpha x^2 + \beta x + \gamma) + b \\ &= a\alpha x^2 + a\beta x + a\gamma + b \end{aligned}$$
which is a quadratic function. ∎

**Example 84** *(Exercise Set 1.11, Exercise 25) A spherical balloon is inflated so that its radius at the end of $t$ seconds is $r(t) = 3\sqrt{t} + 5$ cm, $0 \le t \le 4$. Express the volume and the surface area as a function of time. What are the units of these quantities?*

Solution: The volume of the balloon as a function of the radius is
$$V(r) = \frac{4}{3}\pi r^3.$$
Since $r(t) = 3\sqrt{t} + 5$, the volume as a function of time is the composition
$$\begin{aligned} V(t) &= V(r(t)) \\ &= \frac{4}{3}\pi(3\sqrt{t} + 5)^3 \text{ cm}^3. \end{aligned}$$

The surface area of a balloon as a function of the radius is

$$S(r) = 4\pi r^2$$

and as a function of time is the composition

$$\begin{aligned} S(t) &= S(r(t)) \\ &= 4\pi(3\sqrt{t}+5)^3 \text{ cm}^2. \end{aligned}$$

■

## 1.12 Inverse Functions

The *inverse* function, when it exists (not all functions have an inverse function), is the process that undoes the original operation of the function. For example, the function $f(x) = 2x - 1$ sends the real number $x = 2$ to the real number $f(2) = 3$. The inverse process sends the number 3 back the originating value 2.

### 1.12.1 One-to-One Functions

Functions that have inverses are those functions that are *one-to-one*. A function is one-to-one provided no two different $x$ values are sent to the same $y$ value. This can be written as

$$f(x_1) \neq f(x_2), \text{ whenever } x_1 \neq x_2$$

or equivalently

$$\text{if } f(x_1) = f(x_2), \text{ then } x_1 = x_2.$$

Geometrically, a function will be one-to-one provided any horizontal line that crosses the graph of the function crosses it only one place. For example, $f(x) = 2x - 1$ is one-to-one, but $g(x) = x^2$ is not one-to-one.

## 1.12. INVERSE FUNCTIONS

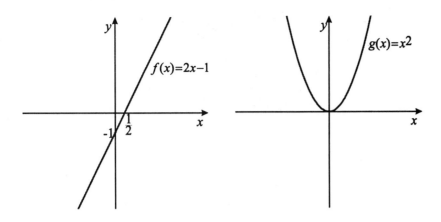

**Example 85** Determine if the function is one-to-one.
(a) $f(x) = x^3 - 1$ (b) (Exercise Set 1.12, Exercise 8) $f(x) = |x|$

Solution:
(a) Use the second equivalent statement to verify the function is one-to-one. If
$$f(x_1) = f(x_2)$$
then
$$x_1^3 - 1 = x_2^3 - 1$$
$$x_1^3 = x_2^3$$
and
$$x_1 = x_2.$$

So the function is one-to-one.
(b) We can determine this function is not one-to-one by inspection. Recall its graph, and note, for example, that
$$f(-2) = |-2| = |2| = f(2).$$

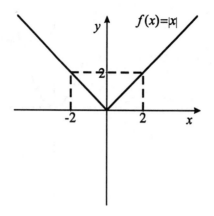

■

## 1.12.2 Process for Finding an Inverse Function

When a function is one-to-one, the following process can often be used to find the inverse. The inverse of the function $f$ is denoted $f^{-1}$.

<u>Procedure for finding the Inverse for $f$</u>
1. Set $y = f(x)$
2. Solve for $x$ in terms of $y$
3. Interchange the variables $x$ and $y$

An important relationship between the graph of a function and its inverse is that they are reflections of each other about the line $y = x$.

**Example 86** *Find the inverse of the one-to-one function. Sketch $y = f(x)$ and $y = f^{-1}(x)$.*
(a) $f(x) = \dfrac{2x - 1}{3}$ (b) $f(x) = \sqrt{x - 1}$

Solution:
(a)
<u>Step 1:</u>
$$y = \frac{2x - 1}{3}$$
<u>Step 2:</u>
$$\begin{aligned} 3y &= 2x - 1 \\ 3y + 1 &= 2x \\ x &= \frac{3y + 1}{2} \end{aligned}$$

## 1.12. INVERSE FUNCTIONS

**Step 3:**
$$f^{-1}(x) = \frac{3x+1}{2}$$

Note that if, for example, $x = -2$, then

$$f(-2) = -\frac{5}{3}$$
$$f^{-1}(f(-2)) = f^{-1}\left(-\frac{5}{3}\right) = \frac{3\left(-\frac{5}{3}\right)+1}{2} = \frac{-4}{2} = -2.$$

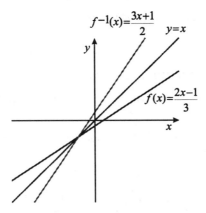

**(b)**
**Step 1:**
$$y = \sqrt{x-1}$$

**Step 2:**
$$y^2 = x - 1$$
$$x = y^2 + 1$$

**Step 3:**
$$f^{-1}(x) = x^2 + 1$$

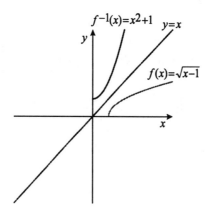

In the figure, the inverse is shown only for $x \geq 0$. This is due to the domain restriction. Another important relationship between a function and its inverse is

the domain of $f$ is the range of $f^{-1}$

the range of $f^{-1}$ is the domain of $f$.

In the example we have the information in the following table.

| Function | Domain | Range |
|---|---|---|
| $f(x) = \sqrt{x-1}$ | $[1, \infty)$ | $[0, \infty)$ |
| $f^{-1}(x) = x^2 + 1$ | $[0, \infty)$ | $[1, \infty)$ |

∎

**Example 87** *(Exercise Set 1.12, Exercise 24) Let $f(x) = x^2 - 2x$. Show that the function is not one-to-one. Determine a subset of the domain of the function on which it is one-to-one, find its inverse on this restricted domain, and specify the domain of the corresponding inverse function.*

Solution: The function can be rewritten as

$$\begin{aligned} f(x) &= x^2 - 2x \\ &= x^2 - 2x + 1 - 1 \\ &= (x-1)^2 - 1. \end{aligned}$$

## 1.12. INVERSE FUNCTIONS

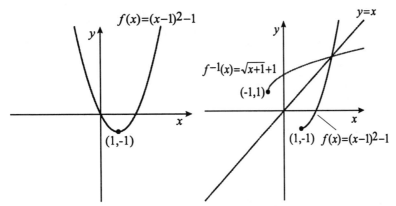

The figure shows that the function does not satisfy the horizontal line test since many horizontal lines cross the graph two times. If the domain is restricted to $[1, \infty)$, then the new function will be one-to-one. The inverse is then found by

Step 1:
$$\begin{aligned} y &= x^2 - 2x \\ &= (x-1)^2 - 1 \end{aligned}$$

Step 2:
$$\begin{aligned} y + 1 &= (x-1)^2 \\ x - 1 &= \sqrt{y+1} \\ x &= \sqrt{y+1} + 1 \end{aligned}$$

Step 3:
$$f^{-1}(x) = \sqrt{x+1} + 1.$$

The domain of $f^{-1}$, which is the same as the range of $f$, is the interval $[-1, \infty)$.
∎

# Chapter 2

# Algebraic Functions

## 2.1 Introduction

A *polynomial function of degree n* is a function of the form

$$P(x) = a_n x^n + a_{n-1} x^{n-1} + \cdots + a_1 x + a_0,$$

where $a_0, a_1, \ldots a_n$ are real numbers and $a_n \neq 0$. The term $a_n$ is called the *leading coefficient* and $a_0$ is called the *constant term*. The *algebraic functions* are obtained from the polynomials by any finite combination of the operations of addition, subtraction, multiplication, division, and extracting integral roots.

## 2.2 Polynomial Functions

**Example 88** *Specify the degree, leading coefficient, and constant term of the polynomial.*
(a) $P(x) = 2x^4 - x^3 + 2x^2 - x + 1$ (b) $P(x) = x^5 - 8$
(c) $P(x) = 3x - 2$ (d) $P(x) = x^{12}$

Solution:

| Polynomial | Degree | Leading Coefficient | Constant Term |
|---|---|---|---|
| $2x^4 - x^3 + 2x^2 - x + 1$ | 4 | 2 | 1 |
| $x^5 - 8$ | 5 | 1 | $-8$ |
| $3x - 2$ | 1 | 3 | $-2$ |
| $x^{12}$ | 12 | 1 | 0 |

## 2.2.1 Graphing Polynomial Functions

**Example 89** *Use the graph of $f(x) = x^3$ shown in the figure to sketch the graph of the function.*
(a) $g(x) = (x-1)^3 + 2$  (b) $g(x) = \frac{1}{2}(x+2)^3 - 1$

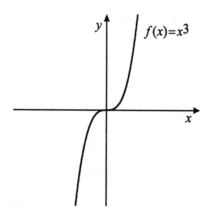

Solution:
(a) Knowing the general shape of the graph of $f(x) = x^3$, the graph of $y = g(x)$ can be obtained quickly using the shifting, scaling, and reflecting techniques. To sketch $y = g(x) = (x-1)^3 + 2$, first sketch $y = (x-1)^3$ by shifting the graph of $y = x^3$ to the right 1 unit, and then shift the resulting curve 2 units upward.

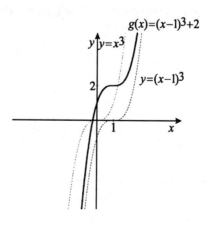

## 2.2. POLYNOMIAL FUNCTIONS

(b) First scale $y = x^3$ vertically by a factor of $\frac{1}{2}$. The new curve is below $y = x^3$ for $x > 0$, and above for $x < 0$. For example, the point $(1, 1)$ is on the graph of $y = x^3$ but the point $\left(1, \frac{1}{2}\right)$ is on the graph of $y = \frac{1}{2}x^3$. Similarly, the point $(-1, -1)$ is on the graph of $y = x^3$ but $\left(-1, -\frac{1}{2}\right)$ is on the graph of $y = \frac{1}{2}x^3$. Now shift $y = \frac{1}{2}x^3$, to the left 2 units, and 1 unit downward to obtain the final graph.

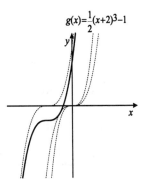

■

**Example 90** *Use the graph of $f(x) = x^4$ shown in the figure to sketch the graph of the curve.*
  *(a) $y = ax^4$, for $a = 2, 3, \frac{1}{2}, \frac{1}{3}$.*
  *(b) $y = ax^4$, for $a = -1, -2, -3, -\frac{1}{2}, -\frac{1}{3}$.*
  *(c) $y = x^4 - a$, for $a = 1, 2, -1, -2$.*
  *(d) $y = (x - a)^4$, for $a = 1, 2, -1, -2$.*

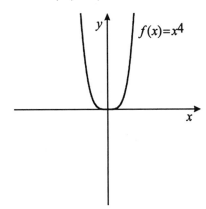

Solution:
(a) If the vertical line $x = 1$ is drawn to intersect the curves $y = ax^4$, for $a = 2, 3, \frac{1}{2}, \frac{1}{3}$, then it crosses the curves at the points shown below.

| Curve | $y = x^4$ | $y = 2x^4$ | $y = 3x^4$ | $y = \frac{1}{2}x^4$ | $y = \frac{1}{3}x^4$ |
|---|---|---|---|---|---|
| Intersects $x = 1$ | $(1,1)$ | $(1,2)$ | $(1,3)$ | $(1,\frac{1}{2})$ | $(1,\frac{1}{3})$ |

Since the curves are symmetric about the $y$-axis the same relationship holds for negative $x$. Using the one point as a model, we can draw the curves in relation to $y = x^4$ as shown in the figure.

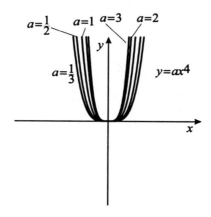

(b) The curves are the reflection of the curves in part (a) about the $x$-axis.

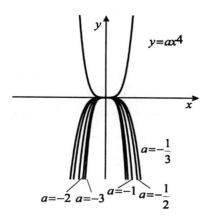

(c) If $a > 0$, then $y = x^4 - a$ is obtained from a vertical shift of $y = x^4$ downward $a$ units. If $a < 0$, then shift upward $|a|$ units.

## 2.2. POLYNOMIAL FUNCTIONS

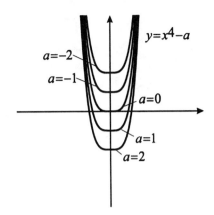

(d) If $a > 0$, then $y = (x-a)^4$ is obtained from a horizontal shift of $y = x^4$ to the right $a$ units. If $a < 0$, then shift to the left $|a|$ units.

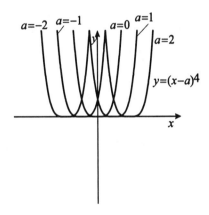

■

### 2.2.2 Zeros and End Behavior in Sketching Polynomials

A number $c$ is called a *zero* of a function $f$ if $f(c) = 0$. If a polynomial $P(x)$ has a factor of the form $(x-c)^k$, then $x = c$ is a *repeated* zero of $P$ of *multiplicity* $k$.

**Example 91** *Find the zeros of the polynomial and specify the multiplicity of any repeated zeros.*
 (a) $P(x) = x^3 - x$  (b) $P(x) = x^3 - 2x^2 + x$

Solution:
(a) To find the zeros of a polynomial factor the polynomial. So,

$$\begin{aligned} P(x) &= x^3 - x \\ &= x(x^2 - 1) \\ &= x(x-1)(x+1). \end{aligned}$$

Then if

$$\begin{aligned} P(x) &= 0 \\ x(x-1)(x+1) &= 0 \Rightarrow \\ x &= 0, \quad x = 1, \quad x = -1. \end{aligned}$$

(b) If

$$\begin{aligned} P(x) &= 0 \\ x^3 - 2x^2 + x &= 0 \\ x(x^2 - 2x + 1) &= 0 \\ x(x-1)(x-1) &= 0 \\ x(x-1)^2 &= 0 \Rightarrow \\ x &= 0, \quad x = 1. \end{aligned}$$

The zero $x = 1$, is of multiplicity 2.
■

The zeros of a polynomial give the $x$-intercepts of the graph of the function. If the multiplicity of a zero is an even number, then the graph flattens and just touches the $x$-axis without crossing the axis. If the multiplicity of the zero is an odd number greater than one, the graph flattens and then crosses the $x$-axis at the zero. This together with the end behavior of the polynomial gives enough information to sketch the graph.

The *end behavior* is what happens to the graph of the polynomial as $|x|$ becomes large, that is, at very large values of $x$ on the positive or negative $x$-axis. The end behavior of a polynomial of degree $n$ depends only on the $x^n$ term. This is written as

$$a_n x^n + a_{n-1} x^{n-1} + \cdots + a_0 \approx a_n x^n.$$

## 2.2. POLYNOMIAL FUNCTIONS

**Example 92** *Test the end behavior of the polynomial.*
(a) $P(x) = x^3 + x^2 + 1$  (b) $P(x) = 5x^4 + 10x^3 + 100x^2 - x + 2$
(c) $P(x) = -3x^5 + x - 1$  (d) $P(x) = -10x^8 + x^7 + 500x^5 + x - 2$

Solution:
(a)
$$P(x) = x^3 + x^2 + 1$$
$$\approx x^3$$

As $|x|$ gets large the graph behaves like the graph $y = x^3$. So, as $x$ goes to $\infty$, $y = P(x) = x^3 + x^2 + 1$ also goes to $\infty$, and as $x$ goes to $-\infty$, $y = P(x) = x^3 + x^2 + 1$ goes to $-\infty$. This is written as

$$x \to \infty \Rightarrow P(x) \to \infty$$
$$x \to -\infty \Rightarrow P(x) \to -\infty.$$

(b)
$$P(x) = 5x^4 + 10x^3 + 100x^2 - x + 2$$
$$\approx 5x^4$$

$$x \to \infty \Rightarrow P(x) \to \infty$$
$$x \to -\infty \Rightarrow P(x) \to \infty$$

(c)
$$P(x) = -3x^5 + x - 1$$
$$\approx -3x^5$$

$$x \to \infty \Rightarrow P(x) \to -\infty$$
$$x \to -\infty \Rightarrow P(x) \to \infty$$

(d)
$$P(x) = -10x^8 + x^7 + 500x^5 + x - 2$$
$$\approx -10x^8$$

$$x \to \infty \Rightarrow P(x) \to -\infty$$
$$x \to -\infty \Rightarrow P(x) \to -\infty$$

■

**Example 93** *Use the zeros of the polynomial and the end behavior to sketch the graph.*
(a) $P(x) = x^3 - x^2 - 2x$  (b) $P(x) = x^3 - 2x^2 + x$
(c) $P(x) = (x-1)(x-2)(x+1)^3$  (d) $P(x) = (x+1)^5(x-1)$

Solution:
(a)
Zeros: Set $P(x) = 0$ and factor.

$$\begin{aligned} x^3 - x^2 - 2x &= 0 \\ x(x^2 - x - 2) &= 0 \\ x(x-2)(x+1) &= 0 \Rightarrow \\ x &= 0, x = 2, x = -1 \end{aligned}$$

End Behavior:

$$\begin{aligned} P(x) &= x^3 - x^2 - 2x \\ &\approx x^3 \end{aligned}$$
$$x \to \infty \Rightarrow P(x) \to \infty$$
$$x \to -\infty \Rightarrow P(x) \to -\infty$$

There is only one way the curve can cross the $x$-axis at $-1, 0,$ and $2$ and satisfy the end behavior. The curve has to come *up* through $x = -1$, rather than down through the point. The curve turns somewhere between $x = -2$ and $x = 0$, passes through $x = 0$ from above, turns again between $x = 0$ and $x = 2$, and goes up through $x = 2$. The curve is shown in the figure. Without the zeros *and* the end behavior we would not be certain how to sketch the curve.

We still cannot determine exactly where the turning points are. This is left to calculus. We can approximate the highs and lows, called *local maximums* and *local minimums*, using a graphing device. For example, if after plotting the curve using a graphing device we click on the local maximum between $x = -1$ and $x = 0$, the point is approximately $(-0.6, 0.6)$.

## 2.2. POLYNOMIAL FUNCTIONS

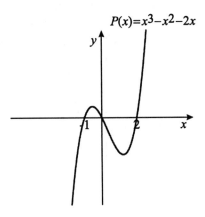

(b)

Zeros: Set $P(x) = 0$ and factor.

$$\begin{align}
x^3 - 2x^2 + x &= 0 \\
x(x^2 - 2x + 1) &= 0 \\
x(x-1)(x-1) &= 0 \\
x(x-1)^2 &= 0 \Rightarrow \\
x &= 0, x = 1
\end{align}$$

End Behavior:

$$\begin{align}
P(x) &= x^3 - 2x^2 + x \\
&\approx x^3 \\
x \to \infty &\Rightarrow P(x) \to \infty \\
x \to -\infty &\Rightarrow P(x) \to -\infty
\end{align}$$

The polynomial has a zero of multiplicity 2 at $x = 1$, so the curve flattens and just touches, without crossing, the $x$-axis at $x = 1$. Applying the end behavior produces the curve as shown in the figure.

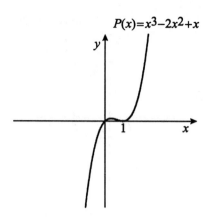

(c)
<u>Zeros:</u> The zeros can be read from the function and are $x = 1$, $x = 2$, and $x = -1$ is a zero of multiplicity 3.

<u>End Behavior:</u> The factors of $P(x)$ are of degree 1, 1, and 3, so if the factors are expanded and terms collected, the degree of the polynomial is 5, and the leading coefficient is 1. Checking,

$$\begin{aligned} P(x) &= (x-1)(x-2)(x+1)^3 \\ &= x^5 - 4x^3 - 2x^2 + 3x + 2 \\ &\approx x^5 \\ x &\to \infty \Rightarrow P(x) \to \infty \\ x &\to -\infty \Rightarrow P(x) \to -\infty. \end{aligned}$$

The curve at the zero $x = -1$ flattens and also crosses the $x$-axis. Putting this information together the curve is as shown in the figure.

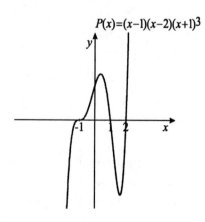

## 2.2. POLYNOMIAL FUNCTIONS

(d)

Zeros: The zeros can be read from the function and are $x = 1$, and $x = -1$ is a zero of multiplicity 5.

End Behavior: The degree of $P(x)$ is 6, and the leading coefficient is 1.

$$\begin{aligned} P(x) &= (x+1)^5(x-1) \\ &= x^6 + 4x^5 + 5x^4 - 5x^2 - 4x - 1 \\ &\approx x^6 \end{aligned}$$

$$x \to \infty \Rightarrow P(x) \to \infty$$
$$x \to -\infty \Rightarrow P(x) \to \infty$$

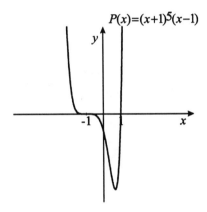

■

**Example 94** *(Exercise Set 2.2, Exercise 16) The cubic polynomial $P(x)$ has a zero of multiplicity one at $x = 2$, a zero of multiplicity two at $x = 1$, and $P(-1) = 4$. Determine $P(x)$ and sketch the graph.*

Solution: A polynomial with the specified zeros is

$$\begin{aligned} Q(x) &= (x-2)(x-1)^2 \\ &= x^3 - 4x^2 + 5x - 2. \end{aligned}$$

But

$$\begin{aligned} Q(-1) &= (-1)^3 - 4(-1)^2 + 5(-1) - 2 \\ &= -12. \end{aligned}$$

The problems asks for a polynomial which when evaluated at $-1$ has a value of 4. So, rather then starting with $Q$, we begin with

$$\begin{aligned} P(x) &= a(x-2)(x-1)^2 \\ &= a(x^3 - 4x^2 + 5x - 2). \end{aligned}$$

If $P(-1) = 4$, then

$$\begin{aligned} 4 &= a(-1-2)(-1-1)^2 \\ 4 &= -12a \\ a &= -\frac{1}{3} \end{aligned}$$

and

$$P(x) = -\frac{1}{3}x^3 + \frac{4}{3}x^2 - \frac{5}{3}x + \frac{2}{3}.$$

The end behavior is

$$\begin{aligned} P(x) &= -\frac{1}{3}x^3 + \frac{4}{3}x^2 - \frac{5}{3}x + \frac{2}{3} \\ &\approx -\frac{1}{3}x^3 \end{aligned}$$

$$x \to \infty \Rightarrow P(x) \to -\infty$$
$$x \to -\infty \Rightarrow P(x) \to \infty.$$

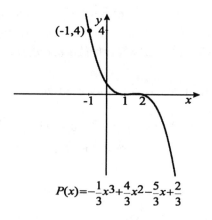

■

## 2.2. POLYNOMIAL FUNCTIONS

### 2.2.3 Applications

**Example 95** *(Exercise Set 2.2, Exercise 27)* *A box without a lid is constructed from a $20'' \times 20''$ piece of cardboard as shown in the figure.*

*(a) Determine the volume of the box as a function of the variable $x$.*

*(b) Use a graphing device to approximate, to within 1 decimal place, the value of $x$ that produces a volume of 500 in$^3$.*

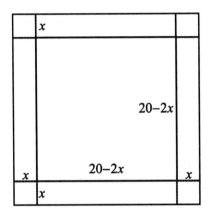

Solution:

(a) If squares of side $x$ inches are removed from the piece of cardboard, as shown in the figure, then the base of the resulting box will have side $20 - 2x$ and height $x$. Then the volume of the box is

$$\begin{aligned} V(x) &= (\text{area of base}) \times (\text{height}) \\ &= \text{length} \times \text{width} \times \text{height} \\ &= (20 - 2x)(20 - 2x)x \\ &= (20 - 2x)^2 x. \end{aligned}$$

(b) To solve $V(x) = (20 - 2x)^2 x = 500$, use a graphing device to plot both $V(x) = (20 - 2x)^2 x$ and the horizontal line $y = 500$ as shown in the figure. There are two values of $x$, for the size of the cutout that will result in a box with volume 500,

$$x \approx 1.9, \text{ and } x \approx 5.$$

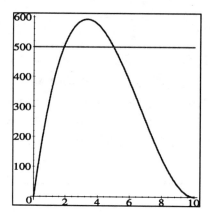

∎

## 2.3 Finding Factors and Zeros of Polynomials

A polynomial $D(x)$ is a *factor* of another polynomial $P(x)$ provided there is a third polynomial $Q(x)$ with

$$P(x) = D(x) \cdot Q(x)$$

In general, $D(x)$ may not divide evenly into $P(x)$, in which case there is a remainder term, say $R(x)$, whose degree is less than $Q(x)$, and

$$P(x) = D(x) \cdot Q(x) + R(x).$$

For this case we can also write

$$\frac{P(x)}{D(x)} = Q(x) + \frac{R(x)}{D(x)}.$$

This is similar to the situation of the long division of positive real numbers. If

$$d \overline{\smash{\big)}\, a} \phantom{x} \begin{matrix} q \\ \vdots \\ \overline{r} \end{matrix}$$

## 2.3. FINDING FACTORS AND ZEROS OF POLYNOMIALS

with $r < d$, then
$$a = dq + r, \text{ or } \frac{a}{d} = q + \frac{r}{d}.$$

For example, if 7 is divided by 2, then $7 = 2 \cdot 3 + 1$, or $\frac{7}{2} = 3 + \frac{1}{2}$. The standard terminology used is

|  | Divisor | Dividend | Quotient | Remainder |
|---|---|---|---|---|
| Real Numbers $\frac{a}{d}$ | $d$ | $a$ | $q$ | $r$ |
| Polynomials $\frac{P(x)}{D(x)}$ | $D(x)$ | $P(x)$ | $Q(x)$ | $R(x)$ |

### 2.3.1 Division of Polynomials

**Example 96** *Divide the polynomial $P(x) = x^3 - 2x^2 + x + 1$ by $x - 1$ and specify the quotient and remainder.*

Solution:

$$\begin{array}{r}
x^2 - x \phantom{xxxxx} \\
x-1 \overline{\smash{\big)} \boxed{x^3} - 2x^2 + x + 1} \\
\underline{x^3 - x^2} \phantom{xxxxxx} \\
\boxed{-x^2} + x \phantom{xx} \\
\underline{-x^2 + x} \phantom{xx} \\
1
\end{array}$$

The quotient is $Q(x) = x^2 - x$, and the remainder is $R(x) = 1$, with
$$\begin{aligned} P(x) &= x^3 - 2x^2 + x + 1 \\ &= (x-1)(x^2 - x) + 1. \end{aligned}$$

The division process ends when the degree of the remainder is less than the degree of the divisor. The key terms in the division process are highlighted in boxes. At each step we determine what has to be multiplied by the highest term in the divisor to yield the term in the box. So for the first step, $x \cdot x^2 = x^3$, and at the second step $x \cdot (-x) = -x^2$. Then we multiply each term of the divisor by the same amount and subtract the result. So, for example, after the first step perform the subtraction

$$\begin{array}{r} x^3 - 2x^2 \\ - \phantom{x} x^3 - x^2 \\ \hline \end{array}$$

The result is $x^3 - 2x^2 - (x^3 - x^2) = x^3 - 2x^2 - x^3 + x^2 = -x^2$.

**Example 97** *(Exercise Set 2.3, Exercise 4)* Let $P(x) = 3x^4 + 2x^3 - x + 2$, and $D(x) = x^2 + 2x - 1$. Find the quotient $Q(x)$, and remainder $R(x)$ when $P(x)$ is divided by $D(x)$.

Solution:

$$
\begin{array}{r}
\phantom{x^2+2x-1\,|\,}3x^2 \; - \; 4x \; + \; 11 \phantom{xxxxxxxxxxxx} \\
x^2 + 2x - 1 \,\overline{\big|\, 3x^4 \; + \; 2x^3 \; + \; 0x^2 \; - \; x \; + \; 2} \\
3x^4 \; + \; 6x^3 \; - \; 3x^2 \phantom{xxxxxxxxxx} \\
\overline{\phantom{xxxx} -\,4x^3 \; + \; 3x^2 \; - \; x \phantom{xxxxx}} \\
-\,4x^3 \; - \; 8x^2 \; + \; 4x \phantom{xxxx} \\
\overline{\phantom{xxxxxxx} 11x^2 \; - \; 5x \; + \; 2} \\
11x^2 \; + \; 22x \; - \; 11 \\
\overline{\phantom{xxxxxxxxxxx} -\,27x \; + \; 13}
\end{array}
$$

The quotient is $Q(x) = 3x^2 - 4x + 11$, the remainder is $R(x) = -27x + 13$, and $3x^4 + 2x^3 - x + 2 = (3x^2 - 4x + 11)(x^2 + 2x - 1) - 27x + 13$.

Inserting a $0x^2$ term aligns each power of $x$ and avoids confusion when subtracting.

∎

### 2.3.2 The Factor Theorem

The Factor Theorem states that if a polynomial $P(x)$ is divided by $x - c$, then the remainder is $P(c)$. In the special case when $P(c) = 0$, we have $x - c$ a factor of $P(x)$ if and only if $c$ is a zero of $P(x)$.

**Example 98** *(Exercise Set 2.3, Exercise 8)* Let $P(x) = 3x^4 + 5x^3 - 5x^2 - 5x + 2$, $c = 1, c = -2$. Use the Factor Theorem to show that $x - c$ is a factor for the given values of $c$, and factor $P(x)$ completely.

Solution: It is not enough to simply verify that $x = 1$ and $x = -2$ are zeros, that is, that $P(1) = 0$ and $P(-2) = 0$. We need to perform long division so we can also factor the polynomial. So

## 2.3. FINDING FACTORS AND ZEROS OF POLYNOMIALS

$$\begin{array}{r}
3x^3 + 8x^2 + 3x - 2 \phantom{)} \\
x-1 \overline{\smash{\big)} 3x^4 + 5x^3 - 5x^2 - 5x + 2} \\
\underline{3x^4 - 3x^3} \phantom{XXXXXXXXXX}\\
8x^3 - 5x^2 \phantom{XXXXXX} \\
\underline{8x^3 - 8x^2} \phantom{XXXXXX} \\
3x^2 - 5x \phantom{XXX} \\
\underline{3x^2 - 3x} \phantom{XXX} \\
-2x + 2 \\
\underline{-2x + 2} \\
0
\end{array}$$

Since the remainder is 0,

$$P(x) = (x-1)(3x^3 + 8x^2 + 3x - 2).$$

To show that $-2$ is a zero of $P(x)$ it is enough to show it is a zero of $3x^3 + 8x^2 + 3x - 2$. After dividing the factor $(x+2)$ into $3x^3 + 8x^2 + 3x - 2$, we have

$$\begin{aligned}
P(x) &= (x-1)(3x^3 + 8x^2 + 3x - 2) \\
&= (x-1)(x+2)(3x^2 + 2x - 1) \\
&= (x-1)(x+2)(3x-1)(x+1).
\end{aligned}$$

■

### 2.3.3 The Rational Zero Test

The rational numbers that are possible zeros of the polynomial

$$P(x) = a_n x^n + a_{n-1} x^{n-1} + \cdots + a_0$$

must be of the form

$$\frac{\pm \text{ factors of } a_0}{\pm \text{ factors of } a_n}.$$

**Example 99** *(Exercise Set 2.3, Exercise 13)* Let

$$P(x) = 10x^5 - 14x^3 + 18x^2 + 6x - 4.$$

Determine all possibilities for rational zeros of $P(x)$.

Solution: The factors of the constant term are $1, 2,$ and $4$ and the factors of the leading coefficient are $1, 2, 5,$ and $10$. So possible rational zeros

$$\frac{\pm 1, \pm 2, \pm 4}{\pm 1, \pm 2, \pm 5, \pm 10} = \pm 1, \pm \frac{1}{2}, \pm \frac{1}{5}, \pm \frac{1}{10}, \pm 2, \pm \frac{2}{5}, \pm 4, \pm \frac{4}{5}.$$

■

Graphing devices are very useful here in eliminating possibilities from the list of possible rational roots, which can often be a very extensive list.

**Example 100** *Find all rational and irrational zeros of the polynomial $P(x) = x^3 - 2x^2 - 5x + 6$, and factor the polynomial completely.*

Solution: The possible rational zeros are

$$\frac{\pm \text{ factors of } 6}{\pm \text{ factors of } 1} = \pm 1, \pm 2, \pm 3, \pm 6.$$

By inspection we can see that $P(1) = 1 - 2 - 5 + 6 = 0$, so $x - 1$ is a factor of the polynomial. This can also be seen from the graph shown in the figure. Then

$$\begin{array}{r}
x^2 \phantom{xx} - \phantom{x} x \phantom{xx} - \phantom{x} 6 \phantom{xx} \\
x-1 \overline{\smash{\big)}\, x^3 \phantom{x} - \phantom{x} 2x^2 \phantom{x} - \phantom{x} 5x \phantom{x} + \phantom{x} 6} \\
\underline{x^3 \phantom{x} - \phantom{x} x^2 \phantom{xxxxxxxxxxxxx}} \\
-x^2 \phantom{x} - \phantom{x} 5x \phantom{xxxxx} \\
\underline{-x^2 \phantom{x} + \phantom{x} x \phantom{xxxxxx}} \\
-6x \phantom{x} + \phantom{x} 6 \\
\underline{-6x \phantom{x} + \phantom{x} 6} \\
0
\end{array}$$

So,

$$\begin{aligned} P(x) &= (x-1)(x^2 - x - 6) \\ &= (x-1)(x-3)(x+2). \end{aligned}$$

## 2.3. FINDING FACTORS AND ZEROS OF POLYNOMIALS

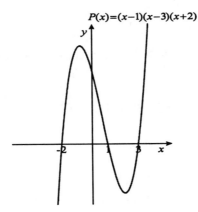

■

**Example 101** *Find all rational and irrational zeros of the polynomial* $P(x) = x^5 - 2x^3 + 2x^2 - 3x + 2$, *and factor the polynomial completely.*

Solution: The possible rational zeros are $\pm 1, \pm 2$. The graph indicates a zero at $x = -2$ and another zero, of multiplicity at least 2, at $x = 1$. If we divide the polynomial by $(x-1)^2$ we get

$$P(x) = (x-1)^2(x^3 + 2x^2 + x + 2).$$

Next divide $x^3 + 2x^2 + x + 2$ by $x + 2$ to get

$$\begin{aligned} P(x) &= (x-1)^2(x^3 + 2x^2 + x + 2) \\ &= (x-1)^2(x+2)(x^2+1). \end{aligned}$$

The polynomial is now in complete factored form since $x^2 + 1$ is never zero and so has no factors.

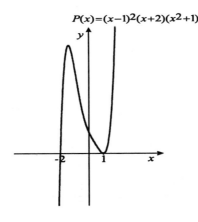

**Example 102** *Find all rational and irrational zeros of the polynomial* $P(x) = 2x^5 - 10x^4 + 7x^3 + 28x^2 - 45x + 18$, *and factor the polynomial completely.*

Solution: The possible rational zeros are

$$\pm 1, \pm 2, \pm 3, \pm 6, \pm 9, \pm 18, \pm\frac{1}{2}, \pm\frac{3}{2}, \pm\frac{9}{2}.$$

The list is extensive, but we could simply substitute each value into the polynomial until we hopefully find a zero, perform long division and continue until the last factor is quadratic which can be handled. A graphing device will reduce the process in this example very quickly.

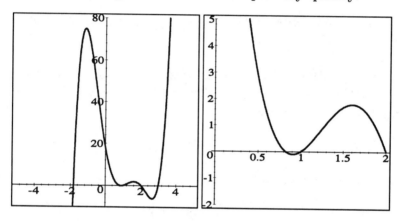

The graph in the figure indicates that there is a zero between $-2$ and $-1$, a zero of multiplicity 2 at $x = 1$ and zeros at $x = 2$ and $x = 3$. The close up view near $x = 1$, indicates $x = 1$ is *not* a root of multiplicity 2! Then

$$(x-1)(x-2)(x-3) = x^3 - 6x^2 + 11x - 6$$

which when divided into $P(x)$ gives

$$P(x) = (x^3 - 6x^2 + 11x - 6)(2x^2 + 2x - 3).$$

To factor the quadratic, we use the quadratic formula. The zeros are

$$x = \frac{-2 \pm \sqrt{4 - 4(2)(-3)}}{4} = \frac{-2 \pm \sqrt{28}}{4}$$
$$= \frac{-2 \pm 2\sqrt{7}}{4} = \frac{-1 \pm \sqrt{7}}{2}.$$

## 2.3. FINDING FACTORS AND ZEROS OF POLYNOMIALS

The final factorization is

$$\begin{aligned} P(x) &= (x^3 - 6x^2 + 11x - 6)(2x^2 + 2x - 3) \\ &= (x-1)(x-2)(x-3)\left(x - \left(\frac{-1+\sqrt{7}}{2}\right)\right)\left(x - \left(\frac{-1-\sqrt{7}}{2}\right)\right). \end{aligned}$$

Note that the Rational Zero test is no help in finding the last two zeros since they are irrational numbers.

■

**Example 103** *(Exercise Set 2.3, Exercise 28) Find a third degree polynomial $P(x)$ that has zeros at $x = -1, x = 1$, and $x = 2$, and whose $x$-term has coefficient 3.*

Solution: A polynomial with the desired zeros is

$$\begin{aligned} Q(x) &= (x+1)(x-1)(x-2) \\ &= x^3 - 2x^2 - x + 2. \end{aligned}$$

Multiplying a polynomial by a constant does not change the zeros of the polynomial, so to make the coefficient of the $x$-term 3, multiply $Q(x)$ by $-3$. So

$$\begin{aligned} P(x) &= -3(x^3 - 2x^2 - x + 2) \\ &= -3x^3 + 6x^2 + 3x - 6. \end{aligned}$$

■

### 2.3.4 Descarte's Rule of Signs

This is a simple test that provides information about the possible *number* of positive real zeros and negative real zeros. If $P(x)$ is a polynomial with real coefficients,

(i) the number of positive zeros is either the number of variations in sign of $P(x)$ or less than this by an even number;

(ii) the number of negative zeros is either the number of variations in sign of $P(-x)$ or less than this by an even number.

**Example 104** *(Exercise Set 2.3, Exercise 18) Use Descarte's Rule of Signs to determine the maximum number of positive and negative zeros of the polynomial* $P(x) = 9x^4 - 9x^3 - 19x^2 + x + 2$.

Solution:
Variations in sign of $P(x)$: $P(x) = \underbrace{9x^4 - 9x^3}_{+\text{ to }-} - \underbrace{19x^2 + x}_{-\text{ to}+} + 2$.

There are two variations in sign, so the maximum number of positive real roots is 2.

Variations in sign of $P(-x)$:

$$\begin{aligned} P(-x) &= 9(-x)^4 - 9(-x)^3 - 19(-x)^2 + (-x) + 2 \\ &= 9x^4 + \underbrace{9x^3 - 19x^2}_{+\text{ to }-} - \underbrace{x + 2}_{-\text{ to }+} \end{aligned}$$

The maximum number of negative real roots is also 2.

∎

## 2.4 Rational Functions

A *rational function* has the form

$$f(x) = \frac{P(x)}{Q(x)}$$

with $P(x)$ and $Q(x)$ polynomials. The domain of $f$ is the set of all real numbers with $Q(x) \neq 0$.

### 2.4.1 Domains and Ranges of Rational Functions

**Example 105** *Determine the domain and range of the rational function whose graph is shown in the figure.*

## 2.4. RATIONAL FUNCTIONS

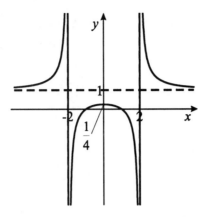

Solution:

Domain: Any vertical line, except $x = -2$, and $x = 1$, will cross the graph of the function in one point, so the domain is $(-\infty, -2) \cup (-2, 2) \cup (2, \infty)$.

Range: Horizontal lines between the line $y = \frac{1}{4}$, and the line $y = 1$, including $y = 1$, will miss the graph and so will be excluded from the range. The range is $(-\infty, \frac{1}{4}] \cup (1, \infty)$.

■

**Example 106** *Find the domain and any x- and y-intercepts of the rational function.*

(a) *(Exercise Set 2.4, Exercise 8)* $f(x) = \dfrac{(x-1)(x+2)}{(x+3)(x-4)}$.

(b) *(Exercise Set 2.4, Exercise 9)* $f(x) = \dfrac{x^2 - x - 6}{x^2 - 4}$.

Solution:

(a) Domain: All real numbers for which the denominator is *not* zero. So

$$(x+3)(x-4) = 0 \Rightarrow$$
$$x = -3, x = 4$$

and the domain is $(-\infty, -3) \cup (-3, 4) \cup (4, \infty)$.

x-intercepts: Solve $y = f(x) = 0$. The fraction will be 0 provided the numerator is 0, so

$$(x-1)(x+2) = 0 \Rightarrow$$
$$x = 1, x = -2$$

and the graph will cross the $x$-axis at the points $(1,0)$ and $(-2,0)$.
   <u>$y$-intercept</u>: Set $x = 0$, so $y = \frac{(-1)(2)}{(3)(-4)} = \frac{1}{6}$ and the graph crosses the $y$-axis at $\left(0, \frac{1}{6}\right)$.
   (b) First factor the numerator and denominator.

$$\begin{aligned} f(x) &= \frac{x^2 - x - 6}{x^2 - 4} \\ &= \frac{(x-3)(x+2)}{(x-2)(x+2)} \\ &= \frac{x-3}{x-2} \end{aligned}$$

<u>Domain</u>:

$$\begin{aligned} (x-2) &= 0 \Rightarrow \\ x &= 2 \end{aligned}$$

so $x = 2$ must be eliminated from the domain. The value $x = -2$ must also be eliminated from the domain since the original equation is not defined at $x = -2$. The domain is $(-\infty, -2) \cup (-2, 2) \cup (2, \infty)$.
   <u>$x$-intercepts</u>: Solve for $y = f(x) = 0$.

$$\begin{aligned} (x-3) &= 0 \\ x &= 3 \end{aligned}$$

and the graph will cross the $x$-axis at the point $(3, 0)$.
   <u>$y$-intercept</u>: Set $x = 0$, so $y = \frac{3}{2}$ and the graph crosses the $y$-axis at $\left(0, \frac{3}{2}\right)$.
■

## 2.4.2 Horizontal and Vertical Asymptotes

An *asymptote* of a graph is a line that the graph approaches. A horizontal line $y = a$ is a *horizontal asymptote* to the graph of $f$ if $f(x) \to a$ as $x \to \infty$ or $x \to -\infty$. The vertical line $x = a$ is a *vertical asymptote* to the graph of $f$, if $f(x) \to \infty$ or $f(x) \to -\infty$ as $x$ approaches $a$ from the left side or the right side.

To find horizontal asymptotes, use the end behavior of the polynomials in the numerator and denominator of the rational function. For example,

$$f(x) = \frac{2x^3 - 2x^2 - 8x + 2}{3x^3 + x^2 - x + 5} \approx \frac{2x^3}{3x^3} = \frac{2}{3}, \text{ for } |x| \text{ large}.$$

## 2.4. RATIONAL FUNCTIONS

This says that as $x$ is selected further from zero on the positive or negative $x$-axis the values of $f(x)$ get closer to $\frac{2}{3}$. So the curve flattens to the horizontal line $y = \frac{2}{3}$.

Vertical asymptotes can only be vertical lines $x = a$, when the denominator of the rational function is 0 at $x = a$. For example,

$$f(x) = \frac{1}{(x-1)(x+2)}$$

has vertical asymptotes $x = 1$ and $x = -2$.

As $x$ gets close to 1 from the right, $x > 1$, so $x - 1 > 0$, but gets close to zero. The factor $(x+2)$ approaches 3, so $(x-1)(x+2) > 0$ and approaches 0, and $f(x) = \frac{1}{(x-1)(x+1)}$ becomes arbitrarily large. This is written

$$x \to 1^+ \Rightarrow f(x) \to \infty.$$

As $x$ gets close to 1 from the left, $x < 1$, so $x - 1 < 0$, but gets close to zero. The factor $(x+2)$ still approaches 3, so $(x-1)(x+2) < 0$ and approaches 0, and $f(x) = \frac{1}{(x-1)(x+1)}$ becomes arbitrarily large in magnitude but negative. This is written

$$x \to 1^- \Rightarrow f(x) \to -\infty$$

The same analysis near the vertical line $x = -2$, gives

$$x \to -2^+ \Rightarrow f(x) \to -\infty$$
$$x \to -2^- \Rightarrow f(x) \to \infty.$$

**Example 107** *Sketch the graph of*

$$f(x) = \frac{x^2 - x - 2}{x - 2},$$

*labeling all horizontal and vertical asymptotes and $x$- and $y$-intercepts.*

Solution: First factor the numerator, so

$$f(x) = \frac{(x-2)(x+1)}{x-2} = x + 1, x \neq 2.$$

The value $x = 2$ is not in the domain of $f$ since it makes the denominator 0, and also can not be included in the final definition even though the fraction simplified nicely.

The graph has no horizontal or vertical asymptotes. The intercepts are $(0, 1)$ and $(-1, 0)$. The graph is the same as the graph of $y = x + 1$, except the point $(-2, -1)$ is removed.

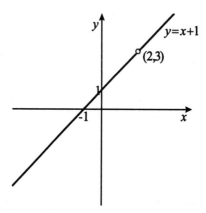

Be careful not to assume a graph has a vertical asymptote whenever the denominator is 0. These are the only possibilities, but if the numerator and denominator have a common factor that can be cancelled, a zero of the denominator may not produce a vertical asymptote.
∎

**Example 108** *(Exercise Set 2.4, Exercise 17) Sketch the graph of*

$$f(x) = \frac{2x - 3}{x^2 - x - 6},$$

*check all horizontal and vertical asymptotes, and x- and y-intercepts.*

Solution: Factor the denominator,

$$\begin{aligned} f(x) &= \frac{2x - 3}{x^2 - x - 6} \\ &= \frac{2x - 3}{(x - 3)(x + 2)}. \end{aligned}$$

<u>Vertical Asymptotes:</u> Set the dominator equal to 0.

$$\begin{aligned} (x - 3)(x + 2) &= 0 \Rightarrow \\ x &= 3, \quad x = -2 \end{aligned}$$

## 2.4. RATIONAL FUNCTIONS

The vertical asymptotes are $x = 3$ and $x = -2$. To sketch the graph, analyze the behavior of the graph near the vertical asymptotes. Whether the graph goes to plus or minus infinity depends on the sign of the fraction.

As $x \to 3^+$, the sign of the fraction is $\dfrac{+}{(+)(+)}$, so $f(x) \to \infty$.

As $x \to 3^-$, the sign of the fraction is $\dfrac{+}{(-)(+)}$, so $f(x) \to -\infty$.

As $x \to -2^+$, the sign of the fraction is $\dfrac{-}{(-)(+)}$, so $f(x) \to \infty$.

As $x \to -2^-$, the sign of the fraction is $\dfrac{-}{(-)(-)}$, so $f(x) \to -\infty$.

<u>Horizontal Asymptotes:</u> The end behavior of the rational function is given by

$$f(x) = \frac{2x-3}{x^2-x-6} \approx \frac{2x}{x^2} \approx \frac{2}{x}$$
$$x \to \infty \Rightarrow f(x) \to 0$$
$$x \to -\infty \Rightarrow f(x) \to 0.$$

So the graph has a horizontal asymptote, $y = 0$.

<u>x-intercepts:</u> Solve

$$f(x) = \frac{2x-3}{x^2-x-6} = 0$$
$$2x - 3 = 0$$
$$2x = 3$$
$$x = \frac{3}{2}.$$

<u>y-intercept:</u> Set $x = 0$, then $y = \frac{1}{2}$.

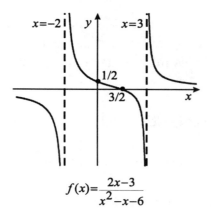

$$f(x) = \frac{2x-3}{x^2-x-6}$$

∎

**Example 109** *(Exercise Set 2.4, Exercise 21) Sketch the graph of*

$$f(x) = \frac{x^2 - 9}{x^2 - 16},$$

*labeling all horizontal and vertical asymptotes, and x- and y-intercepts.*

Solution:

$$\begin{aligned} f(x) &= \frac{x^2 - 9}{x^2 - 16} \\ &= \frac{(x-3)(x+3)}{(x-4)(x+4)} \end{aligned}$$

<u>Vertical Asymptotes:</u> $x = -4, x = 4$

|  | Sign of fraction | Behavior near asymptote |
|---|---|---|
| $x \to -4^+$ | $\frac{(-)(-)}{(-)(+)}$ | $f(x) \to -\infty$ |
| $x \to -4^-$ | $\frac{(-)(-)}{(-)(-)}$ | $f(x) \to \infty$ |
| $x \to 4^+$ | $\frac{(+)(+)}{(+)(+)}$ | $f(x) \to \infty$ |
| $x \to 4^-$ | $\frac{(+)(+)}{(-)(+)}$ | $f(x) \to -\infty$ |

<u>Horizontal Asymptote:</u> $y = 1$

$$f(x) = \frac{x^2 - 9}{x^2 - 16} \approx \frac{x^2}{x^2} = 1$$

## 2.4. RATIONAL FUNCTIONS

x-intercepts: $(3,0), (-3,0)$

$$\frac{(x-3)(x+3)}{(x-4)(x+4)} = 0$$
$$(x-3)(x+3) = 0 \Rightarrow$$
$$x = 3, x = -3$$

y-intercept: $\left(0, \frac{9}{16}\right)$

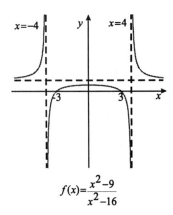

$f(x) = \frac{x^2-9}{x^2-16}$

■

### 2.4.3 Slant Asymptotes

If the degree of the denominator of a rational function is exactly one greater than the degree of the numerator, then the graph will approach a non-horizontal line as $x \to \infty$ and as $x \to -\infty$. Such a line is called a *slant asymptote* to the graph. The equation of the slant asymptote comes from the fact that in this case

$$f(x) = \frac{P(x)}{Q(x)}$$
$$= (ax+b) + \frac{R(x)}{Q(x)}$$

where the degree of $R(x)$ is less than the degree of $Q(x)$. The quotient has the form $ax+b$, since the degree of $P(x)$ is exactly one greater than the degree of $Q(x)$.

**Example 110** *(Exercise Set 2.4, Exercise 27)* Use a graphing device to sketch the graph of
$$f(x) = \frac{2x^2 + 3x - 1}{x + 2}$$
and show any vertical and slant asymptotes, and $x$- and $y$-intercepts.

Solution: There is a vertical asymptote at $x = -2$, and since the degree of the numerator is one greater than the degree of the denominator we have a slant asymptote.

$$\begin{array}{r}
2x \phantom{xx} - \phantom{x} 1 \phantom{xx} \\
x+2 \overline{\smash{)}\, 2x^2 + 3x - 1} \\
\underline{2x^2 + 4x \phantom{xxxx}} \\
-x - 1 \\
\underline{-x - 2} \\
1
\end{array}$$

gives
$$f(x) = 2x - 1 + \frac{1}{x+2}$$

and

$$x \to \infty \Rightarrow \frac{1}{x+2} \to 0 \Rightarrow f(x) \to 2x - 1$$

$$x \to -\infty \Rightarrow \frac{1}{x+2} \to 0 \Rightarrow f(x) \to 2x - 1.$$

The slant asymptote is $y = 2x - 1$.

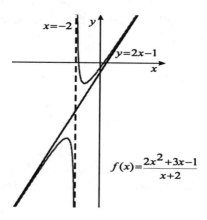

## 2.4. RATIONAL FUNCTIONS

**Example 111** *(Exercise Set 2.4, Exercise 32) Define a rational function that satisfies all of the following conditions:*
  *i) has the vertical asymptotes $x = 2$ and $x = -3$*
  *ii) has the horizontal asymptote $y = 1$*
  *iii) has $x$-intercepts at 3 and 4.*

Solution: For the rational function to have vertical asymptotes $x = 2$ and $x = -3$, the simplified form of the function must have a denominator that contains the factors $(x-2)$ and $(x+3)$. If the graph is to have the horizontal asymptote $y = 1$, the numerator and the denominator must have the same degree and the leading coefficients need to be the same. If the graph is to cross the $x$-axis at 3 and 4, then the numerator must contain the factors $x-3$ and $x-4$. A possible function is

$$f(x) = \frac{(x-3)(x-4)}{(x-2)(x+3)}$$
$$= \frac{x^2 - 7x + 12}{x^2 + x - 6}.$$

### 2.4.4 Application to Optimization

**Example 112** *(Exercise Set 2.4, Exercise 35) A rectangular box with a square base of length $x$ and height $h$ is to have a volume of 20 ft$^3$. The cost of the top and the bottom of the box is 20 cents per square foot and the cost for the sides is 8 cents per square foot. Express the cost of the box in terms of*
  *(a) the variables $x$ and $h$, (b) the variable $x$ only, (c) the variable $h$ only.*
  *(d) Use a graphing device to approximate the dimensions of the box that will minimize the cost.*

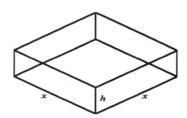

Solution: The cost of a piece of material is the area of the piece times the price per square foot. The surface area of the entire box is

$$\begin{aligned} A &= \text{(area of the top and bottom)} + \text{(area of the four sides)} \\ &= 2x^2 + 4xh. \end{aligned}$$

(a) The cost is then

$$\begin{aligned} C &= (2x^2)(0.20) + (4xh)(0.08) \\ &= 0.4x^2 + 0.32xh. \end{aligned}$$

(b) To eliminate the variable $h$ from the equation for cost, we need an equation relating $h$ and $x$. We have yet to use the information that the volume of the box is 20. The volume is

$$\begin{aligned} V &= \text{(area of the base)} \times \text{(height)} \\ &= x^2 h = 20, \end{aligned}$$

so

$$h = \frac{20}{x^2}.$$

Then

$$\begin{aligned} C(x) &= 0.4x^2 + 0.32xh \\ &= 0.4x^2 + 0.32x\left(\frac{20}{x^2}\right) \\ &= 0.4x^2 + \frac{6.4}{x}. \end{aligned}$$

(c) Use the volume to solve for $x$ in terms of $h$,

$$\begin{aligned} x^2 h &= 20 \\ x^2 &= \frac{20}{h} \\ x &= \sqrt{\frac{20}{h}}. \end{aligned}$$

## 2.5. OTHER ALGEBRAIC FUNCTIONS

So
$$C(h) = 0.4\left(\frac{20}{h}\right) + 0.32\sqrt{\frac{20}{h}}h$$
$$= \frac{8}{h} + 0.32\sqrt{20}\sqrt{h}$$
$$= \frac{8}{h} + 0.64\sqrt{5}\sqrt{h}.$$

(d) The figure shows the graph of $y = C(x) = 0.4x^2 + \frac{6.4}{x}$. The dimension of the side of the base that yields a volume of 20 and uses the smallest amount of material is given by the minimum point on the graph. This occurs at approximately
$$x \approx 2 \Rightarrow$$
$$h = \frac{20}{x^2} \approx 5.$$

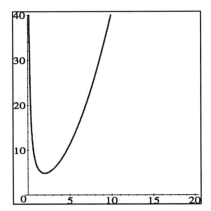

■

## 2.5 Other Algebraic Functions

### 2.5.1 Power Functions

The *rational power functions* have the form
$$f(x) = x^{\frac{m}{n}}$$
$$= \left(\sqrt[n]{x}\right)^m$$
$$= \sqrt[n]{x^m},$$

where $\frac{m}{n}$ is a rational number, and $n$ is an integer greater than 1.

**Example 113** *Use the graph of $y = x^{\frac{1}{3}}$, to sketch the graph of the function.*
(a) $f(x) = (x-2)^{\frac{1}{3}} - 1$  (b) $f(x) = -2(x-1)^{\frac{1}{3}} + 2$
(c) $f(x) = (x+1)^{\frac{2}{3}}$

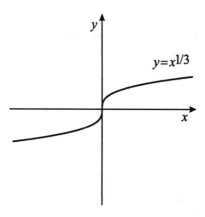

Solution:
(a) Shift the graph of $y = x^{\frac{1}{3}}$, to the right 2 units, and 1 unit downward.

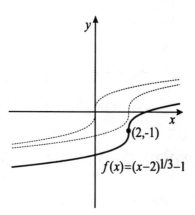

(b) Vertically scale the graph of $y = x^{\frac{1}{3}}$ by a factor of 2, so for $x > 0$ the graph of $y = 2x^{\frac{1}{3}}$ is above the graph of $y = x^{\frac{1}{3}}$, and below it for $x < 0$. Then reflect the resulting graph about the $x$-axis, shift it to the right 1 unit, and upward 2 units to obtain the graph of $f(x) = -2(x-1)^{\frac{1}{3}} + 2$.

## 2.5. OTHER ALGEBRAIC FUNCTIONS

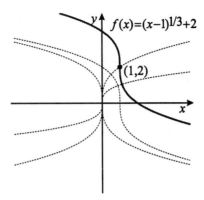

(c) To sketch $y = x^{\frac{2}{3}}$ notice that

$$\text{for } 0 < x < 1, \text{ we have } x^{\frac{2}{3}} = \left(x^{\frac{1}{3}}\right)^2 < x^{\frac{1}{3}}$$

$$\text{for } x > 1, \text{ we have } x^{\frac{2}{3}} = \left(x^{\frac{1}{3}}\right)^2 > x^{\frac{1}{3}}$$

$$\text{for } x = 1, \text{ we have } x^{\frac{2}{3}} = x^{\frac{1}{3}}.$$

Since

$$\begin{aligned}
(-x)^{\frac{2}{3}} &= \left(\sqrt[3]{-x}\right)^2 \\
&= \left(-\sqrt[3]{x}\right)^2 \\
&= \left(\sqrt[3]{x}\right)^2 \\
&= x^{\frac{2}{3}},
\end{aligned}$$

the function $g(x) = x^{\frac{2}{3}}$ is an even function and the graph is symmetric with respect to the $y$-axis. The graph of $f(x) = (x+1)^{\frac{2}{3}}$ is obtained by shifting the graph of $y = x^{\frac{2}{3}}$ to the left 1 unit.

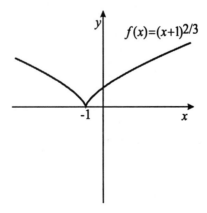

**Example 114** *(Exercise Set 2.5, Exercise 3) Determine the domain of the function* $f(x) = \sqrt{\dfrac{x-1}{x+3}}$.

Solution: The domain is the set of all real numbers that make the expression under the radical greater than or equal to 0. So solve

$$\frac{x-1}{x+3} \geq 0,$$

which is greater than 0 when numerator and denominator are both positive, or are both negative. It is zero when $x - 1 = 0$, or $x = 1$. The linear factors separate the real line into the intervals $(-\infty, -3) \cup (-3, 1) \cup (1, \infty)$. So,

| Interval | Test value | Test | Inequality |
|---|---|---|---|
| $(-\infty, -3)$ | $-4$ | $\frac{-4-1}{-4+3} = 5$ | $> 0$ |
| $(-3, 1)$ | $0$ | $\frac{0-1}{0+3} = -\frac{1}{3}$ | $< 0$ |
| $(1, \infty)$ | $2$ | $\frac{2-1}{2+3} = \frac{1}{5}$ | $> 0$ |

The domain is $(-\infty, -3) \cup [1, \infty)$.

■

**Example 115** *(Exercise Set 2.5, Exercise 14) Sketch the graph of* $f(x) = \sqrt{\dfrac{x-1}{x+2}}$, *showing the x-intercepts and any asymptotes.*

Solution:
**Domain:** $\{x | \frac{x-1}{x+2} \geq 0\} = (-\infty, -2) \cup [1, \infty)$.
**x-intercepts:** Solve

$$\sqrt{\frac{x-1}{x+2}} = 0$$
$$\frac{x-1}{x+2} = 0$$
$$x - 1 = 0$$
$$x = 1.$$

The graph crosses the x-axis at $(1, 0)$.

y-intercepts: None, since $x = 0$ is not in the domain of the function.
Horizontal asymptote: $y = 1$, since for $|x|$ large

$$\sqrt{\frac{x-1}{x+2}} \approx \frac{\sqrt{x}}{\sqrt{x}} = 1.$$

Vertical asymptote: $x = -2$. Because of the domain of $f$, $x$ can only approach $-2$ from the left. So

$$x \to -2^- \Rightarrow f(x) \to \infty.$$

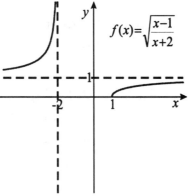

■

## 2.6 Complex Roots of Polynomials

A *complex number* is an expression of the form $a + bi$, where $a$ and $b$ are real numbers and $i = \sqrt{-1}$, that is, $i^2 = -1$. The number $a$ is called the *real part* and $b$ is called the *imaginary part*. The *conjugate* of the complex number $a + bi$ is $\overline{a + bi} = a - bi$.

### 2.6.1 Arithmetic Operations on Complex Numbers

**Example 116** *Write the complex number in standard form $a + bi$.*
  (a) $(1 - 3i) + (2 + 4i)$   (b) $(3 - 2i) - (7 + 9i)$
  (c) $(2 - 3i) \cdot (3 + 5i)$   (d) $(4 - 3i) \cdot (7 - 2i)$
  (e) $\dfrac{2}{3 - i}$   (f) $\dfrac{4 - i}{2 + 3i}$

Solution: Addition and subtraction of complex numbers is performed simply by adding and subtracting real and imaginary parts separately.

(a)
$$(1 - 3i) + (2 + 4i) = (1 + 2) + (-3 + 4)i$$
$$= 3 + i$$

(b)
$$(3 - 2i) - (7 + 9i) = (3 - 7) + (-2 - 9)i$$
$$= -4 - 11i$$

Multiplication is performed using the distributive law for multiplying real numbers, keeping in mind that $i^2 = -1$.

(c)
$$(2 - 3i) \cdot (3 + 5i) = 2(3 + 5i) - 3i(3 + 5i)$$
$$= 6 + 10i - 9i - 15i^2$$
$$= 6 + i - 15(-1)$$
$$= 21 + i$$

(d)
$$(4 - 3i) \cdot \overline{(7 - 2i)} = (4 - 3i) \cdot (7 + 2i)$$
$$= 28 + 8i - 21i - 6i^2$$
$$= 28 - 13i + 6$$
$$= 34 - 13i$$

Complex division is similar to rationalizing a fraction containing radicals. Here we multiply the numerator and denominator of the complex fraction by the conjugate of the denominator.

(e)
$$\frac{2}{3 - i} = \frac{2}{3 - i} \cdot \frac{\overline{3 - i}}{\overline{3 - i}}$$
$$= \frac{2}{3 - i} \cdot \frac{3 + i}{3 + i}$$
$$= \frac{6 + 2i}{9 - i^2}$$

## 2.6. COMPLEX ROOTS OF POLYNOMIALS

$$= \frac{6+2i}{10}$$
$$= \frac{3}{5} + \frac{1}{5}i$$

(f)
$$\frac{4-i}{2+3i} = \frac{4-i}{2+3i} \cdot \frac{2-3i}{2-3i}$$
$$= \frac{8-12i-2i+3i^2}{4-9i^2}$$
$$= \frac{5-14i}{13}$$
$$= \frac{5}{13} - \frac{14}{13}i$$

■

**Example 117** *Show that $1-i$ is a solution to the equation $x^2 - 2x + 2 = 0$.*

Solution: Substituting,

$$\begin{aligned}(1-i)^2 - 2(1-i) + 2 &= (1-i)(1-i) - 2 + 2i + 2 \\ &= 1 - 2i + i^2 + 2i \\ &= 1 + i^2 \\ &= 1 - 1 \\ &= 0.\end{aligned}$$

■

### 2.6.2 Complex Zeros of Polynomials

**Example 118** *(Exercise Set 2.6, Exercise 21) Find the zeros of the quadratic function $f(x) = 2x^2 - x + 2$, and write the function in factored form.*

Solution: Factoring directly is not possible, so we use the quadratic formula to find the zeros.
$$2x^2 - x + 2 = 0$$
has

$$a = 2, b = -1, c = 2$$

so

$$\begin{aligned} x &= \frac{-b \pm \sqrt{b^2 - 4ac}}{2a} \\ &= \frac{1 \pm \sqrt{1 - 4(2)(2)}}{2(2)} \\ &= \frac{1 \pm \sqrt{-15}}{4}. \end{aligned}$$

The discriminant, which is the part under the radical, is negative, so the zeros are complex numbers. Since

$$\sqrt{-15} = \sqrt{15}\sqrt{-1} = \sqrt{15}i,$$

the zeros are

$$x = \frac{1 \pm \sqrt{15}i}{4}.$$

The quadratic can then be factored as

$$\begin{aligned} f(x) &= 2x^2 - x + 2 \\ &= \left(x - \left(\frac{1 + \sqrt{15}i}{4}\right)\right)\left(x - \left(\frac{1 - \sqrt{15}i}{4}\right)\right). \end{aligned}$$

■

Polynomials with real coefficients have the property that if $a + bi$ is a zero, then so is its conjugate $a - bi$.

**Example 119** *(Exercise Set 2.6, Exercise 24) Show that $x = -2 + 4i$ is a solution of the equation $x^3 + 3x^2 + 16x - 20 = 0$ and then find all solutions.*

Solution:

$$\begin{aligned} &(-2 + 4i)^3 + 3(-2 + 4i)^2 + 16(-2 + 4i) - 20 \\ &= (-2 + 4i)\left((-2 + 4i)^2 + 3(-2 + 4i) + 16\right) - 20 \end{aligned}$$

## 2.6. COMPLEX ROOTS OF POLYNOMIALS

$$
\begin{aligned}
&= (-2+4i)((4-16i+16i^2)-6+12i+16)-20 \\
&= (-2+4i)(-12-16i+12i+10)-20 \\
&= (-2+4i)(-2-4i)-20 \\
&= 4-16i^2-20 \\
&= 4+16-20 \\
&= 0
\end{aligned}
$$

Since $-2+4i$ is a zero of the function $f(x) = x^3 + 3x^2 + 16x - 20$, its conjugate $-2-4i$ is also a zero. To find the last solution, divide the polynomial

$$(x-(-2+4i))(x-(-2-4i)) = x^2 + 4x + 20$$

into $x^3 + 3x^2 + 16x - 20$, which must divide $f(x)$ evenly.

$$
\begin{array}{r}
x - 1 \phantom{xxxxxxxxxxx} \\
x^2 + 4x + 20 \overline{\smash{\big)}\, x^3 + 3x^2 + 16x - 20} \\
\underline{x^3 + 4x^2 + 20x\phantom{xxxx}} \\
-x^2 - 4x - 20 \\
\underline{-x^2 - 4x - 20} \\
0
\end{array}
$$

Then

$$x^3 + 3x^2 + 16x - 20 = (x-(-2+4i))(x-(-2-4i))(x-1)$$

and the solutions to $x^3 + 3x^2 + 16x - 20 = 0$ are

$$x = -2+4i, \quad x = -2-4i, \quad x = 1.$$

■

**Example 120** *(Exercise Set 2.6, Exercise 29)* Find all zeros and factor completely the function $f(x) = x^4 - x^2 - 2x + 2$.

Solution:
Possible rational zeros: $\pm 1, \pm 2$
The graph of the function, obtained from a graphing device, appears to indicate a zero of multiplicity 2 at $x = 1$. Dividing $(x-1)^2 = x^2 - 2x + 1$

into $x^4 - x^2 - 2x + 2$ yields a quotient of $x^2 + 2x + 2$ and a remainder of 0. So,
$$f(x) = (x-1)^2(x^2 + 2x + 2).$$
The solutions to $x^2 + 2x + 2 = 0$ are
$$\begin{aligned} x &= \frac{-2 \pm \sqrt{4 - 4(1)(2)}}{2} \\ &= \frac{-2 \pm \sqrt{-4}}{2} \\ &= \frac{-2 \pm 2i}{2} \\ &= -1 \pm i. \end{aligned}$$

<u>Zeros:</u> $x = 1$, of multiplicity 2, $x = -1 + i$, $\quad x = -1 - i$
<u>Factored $f$:</u> $f(x) = (x-1)^2(x - (-1+i))(x - (-1-i))$

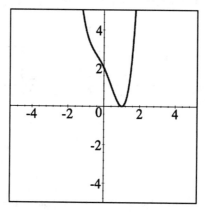

■

**Example 121** *(Exercise Set 2.6, Exercise 34)* Find a polynomial with integer coefficients that has degree 5, zeros $i$, and $3 - i$, and the graph passes through the origin.

Solution: Since the polynomial has real coefficients and $i$ and $3 - i$ are zeros, their conjugates $-i$ and $3 + i$ are also zeros. Since the graph passes through the origin, 0 is another zero. So define the polynomial by

$$\begin{aligned} f(x) &= x(x+i)(x-i)(x-(3-i))(x-(3+i)) \\ &= x^5 - 6x^4 + 11x^3 - 6x^2 + 10x. \end{aligned}$$

■

# Chapter 3

# Trigonometric Functions

## 3.1 Introduction

The *trigonometric functions* $f(t) = \cos t$, and $g(t) = \sin t$ are defined as the $x$- and $y$-coordinates of points $P(t) = (\cos t, \sin t)$ on the unit circle. The point $P(t)$ is called the terminal point and $t$ is the length of the arc along the circle measured counterclockwise from the starting point $(1, 0)$. The measure of $t$ is called *radian measure*. The sine and cosine functions are the basis for the other trigonometric functions. Master these and you have mastered most of trigonometry.

## 3.2 The Sine and Cosine Functions

### 3.2.1 The Unit Circle

The *unit circle* is the circle with center at the origin $(0, 0)$ and radius 1. The equation of the unit circle is

$$x^2 + y^2 = 1.$$

**Example 122** *Verify that the point is on the unit circle.*
(a) $\left(\frac{\sqrt{3}}{2}, \frac{1}{2}\right)$ (b) $\left(-\frac{1}{3}, \frac{2\sqrt{2}}{3}\right)$

Solution:

(a) Substitute the coordinates into the equation for the unit circle and if the equation is satisfied the point will lie on the unit circle. Substituting $x = \frac{\sqrt{3}}{2}$ and $y = \frac{1}{2}$,

$$\left(\frac{\sqrt{3}}{2}\right)^2 + \left(\frac{1}{2}\right)^2 = \frac{(\sqrt{3})^2}{2^2} + \frac{1}{2^2}$$
$$= \frac{3}{4} + \frac{1}{4} = 1$$

and the point lies on the unit circle.

(b) Likewise

$$\left(-\frac{1}{3}\right)^2 + \left(\frac{2\sqrt{2}}{3}\right)^2 = \frac{1}{9} + \frac{2^2(\sqrt{2})^2}{9}$$
$$= \frac{1}{9} + \frac{8}{9} = 1.$$

∎

**Example 123** *Find the point on the unit circle that satisfies the given conditions.*

*(a) x-coordinate $\frac{4}{5}$ and y-coordinate negative.*
*(b) y-coordinate $-\frac{5}{6}$ and in quadrant IV.*
*(c) x-coordinate $-\frac{\sqrt{2}}{2}$, and in quadrant II.*

Solution:
(a) Using the fact that the point lies on the unit circle, and so satisfies the equation $x^2 + y^2 = 1$,

$$\left(\frac{4}{5}\right)^2 + y^2 = 1$$
$$y^2 = 1 - \frac{16}{25} = \frac{9}{25}$$
$$y = \pm\sqrt{\frac{9}{25}} = \pm\frac{3}{5}.$$

Since the y-coordinate is negative we have $\left(\frac{4}{5}, -\frac{3}{5}\right)$.

## 3.2. THE SINE AND COSINE FUNCTIONS

(b) Solving for the $x$-coordinate,

$$x^2 + \left(-\frac{5}{6}\right)^2 = 1$$

$$x^2 = 1 - \frac{25}{36} = \frac{11}{36}$$

$$x = \pm\frac{\sqrt{11}}{6}.$$

The quadrants of the plane are labeled I, II, III and IV counterclockwise starting in the upper right. In quadrant IV the $x$-coordinate is positive and $y$-coordinate is negative, so

$$x = \frac{\sqrt{11}}{6}$$

and the point is $\left(\frac{\sqrt{11}}{6}, -\frac{5}{6}\right)$.

(c)

$$\left(-\frac{\sqrt{2}}{2}\right)^2 + y^2 = 1$$

$$y^2 = 1 - \frac{1}{2} = \frac{1}{2}$$

$$y = \pm\sqrt{\frac{1}{2}} = \pm\frac{1}{\sqrt{2}} = \pm\frac{1}{\sqrt{2}} \cdot \frac{\sqrt{2}}{\sqrt{2}}$$

$$y = \pm\frac{\sqrt{2}}{2}$$

In quadrant II the $y$-coordinate is positive, so we have $\left(-\frac{\sqrt{2}}{2}, \frac{\sqrt{2}}{2}\right)$.

■

### 3.2.2 Terminal Points

If $t > 0$, then $P(t)$ is the point on the unit circle found by measuring $t$ units counterclockwise along the arc of the circle starting at $(1, 0)$. If $t < 0$, then measure clockwise around the circle starting at $(1, 0)$.

**Example 124** *Find the location of the terminal point $P(t)$ on the unit circle for the given value of $t$.*

(a) $t = \frac{\pi}{4}$  (b) $t = \frac{7\pi}{6}$  (c) $t = 7\pi$  (d) $t = -\frac{3\pi}{2}$  (e) $t = -\frac{4\pi}{3}$

Solution:

(a) The circumference of the unit circle is $2\pi$, so $\frac{\pi}{2}$ is one fourth the way around. Since $\frac{\pi}{4} = \frac{1}{2}(\frac{\pi}{2})$, so $\frac{\pi}{4}$ is one eighth the way around the circle.

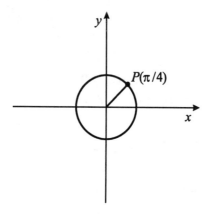

(b) Since $\frac{\pi}{6} = \frac{1}{3}(\frac{\pi}{2})$, an arc of length $\frac{\pi}{6}$ is one third of the arc from $(1,0)$ to $(0,1)$. Also $P(\pi) = (-1,0)$ since an arc of $\pi$ units is half the unit circle. So $P\left(\frac{7\pi}{6}\right)$ is one third of the way between $(-1,0)$ and $(0,-1)$.

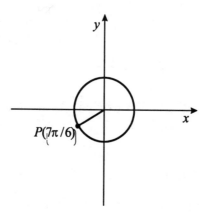

(c) An arc of length $\pi$ is half the length of the unit circle, or one half a complete revolution. $2\pi$ is one revolution around the circle, $4\pi$ is two revolutions, $6\pi$ three revolutions, $7\pi$ three and one half revolutions, and $P(7\pi)$ is located at the point $(-1,0)$.

## 3.2. THE SINE AND COSINE FUNCTIONS

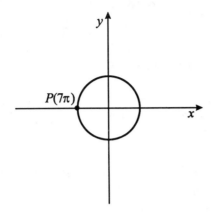

(d) $P\left(-\frac{3\pi}{2}\right)$ is three fourths of the way around the circle in the clockwise direction and is located at $(0,1)$.

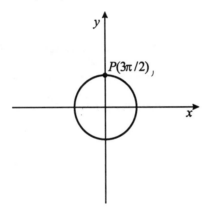

(e) Since $-\frac{\pi}{3} = \frac{2}{3}\left(-\frac{\pi}{2}\right)$, the point $P\left(-\frac{\pi}{3}\right)$ is two thirds of the way from $(1,0)$ to $(0,-1)$. Also $-\pi = 3\left(-\frac{\pi}{3}\right)$, so $P\left(-\frac{4\pi}{3}\right)$ is two thirds of the way between $(-1,0)$ and $(0,1)$.

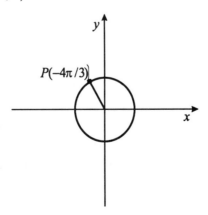

### 3.2.3 Reference Number

For any real number $t$ the *reference number* $r$ associated with $t$ is the shortest distance along the unit circle from $t$ to the $x$-axis. For any $t$ the reference number satisfies $0 \leq t \leq \frac{\pi}{2}$.

**Example 125** *Find the reference number $r$ for the given value of $t$ and show $P(t)$ and $P(r)$ on the unit circle.*
(a) $t = \frac{5\pi}{6}$ (b) $t = -\frac{2\pi}{3}$ (c) $t = \frac{5\pi}{4}$

Solution:

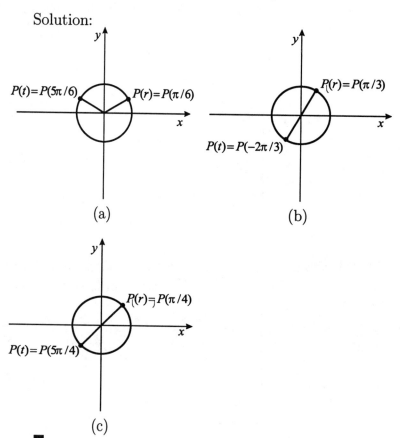

(a)  (b)

(c)

**Example 126** *(Exercise Set 3.2, Exercise 31) If $P(t)$ has coordinates $\left(\frac{3}{5}, \frac{4}{5}\right)$, find the coordinates of*
(a) $P(t + \pi)$ (b) $P(-t)$ (c) $P(t - \pi)$ (d) $P(-t - \pi)$

## 3.2. THE SINE AND COSINE FUNCTIONS

Solution:

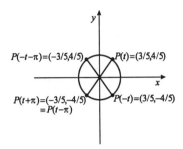

### 3.2.4 Sine and Cosine

If a point on the unit circle has coordinates $P(t) = (x, y)$, then

$$\sin t = y, \quad \cos t = x.$$

These definitions are also used for the sine and cosine of an angle with radian measure $t$. The information in the following table is basic.

| $t$ | $0$ | $\frac{\pi}{6}$ | $\frac{\pi}{4}$ | $\frac{\pi}{3}$ | $\frac{\pi}{2}$ | $\pi$ | $\frac{3\pi}{2}$ | $2\pi$ |
|---|---|---|---|---|---|---|---|---|
| $P(t)$ | $(1,0)$ | $\left(\frac{\sqrt{3}}{2}, \frac{1}{2}\right)$ | $\left(\frac{\sqrt{2}}{2}, \frac{\sqrt{2}}{2}\right)$ | $\left(\frac{1}{2}, \frac{\sqrt{3}}{2}\right)$ | $(0,1)$ | $(-1,0)$ | $(0,-1)$ | $(1,0)$ |
| $\sin t$ | $0$ | $\frac{1}{2}$ | $\frac{\sqrt{2}}{2}$ | $\frac{\sqrt{3}}{2}$ | $1$ | $0$ | $-1$ | $0$ |
| $\cos t$ | $1$ | $\frac{\sqrt{3}}{2}$ | $\frac{\sqrt{2}}{2}$ | $\frac{1}{2}$ | $0$ | $-1$ | $0$ | $1$ |

**Example 127** *Find* $\sin t$ *and* $\cos t$ *for the given value of* $t$.
(a) $t = \dfrac{3\pi}{4}$  (b) $t = \dfrac{7\pi}{6}$  (c) $t = \dfrac{5\pi}{3}$  (d) $t = -\dfrac{7\pi}{4}$

Solution:
(a) The reference number for $t$ is $r = \frac{\pi}{4}$. To find the sine and cosine of $t$, use the sine and cosine of $r$ and determine the appropriate sign for each based on the quadrant where $P(t)$ lies. So

$$\sin \frac{\pi}{4} = \frac{\sqrt{2}}{2}, \quad \cos \frac{\pi}{4} = \frac{\sqrt{2}}{2}$$

and since $P\left(\frac{3\pi}{4}\right)$ lies in quadrant II, $\sin \frac{3\pi}{4} > 0$ and $\cos \frac{3\pi}{4} < 0$, and

$$\sin\frac{3\pi}{4} = \frac{\sqrt{2}}{2}, \qquad \cos\frac{3\pi}{4} = -\frac{\sqrt{2}}{2}.$$

(b)
Reference number: $r = \frac{\pi}{6}$
Quadrant of $P(t)$ : III, so $\sin t < 0$ and $\cos t < 0$
So

$$\sin\frac{7\pi}{6} = -\sin\frac{\pi}{6} = -\frac{1}{2}, \qquad \cos\frac{7\pi}{6} = -\cos\frac{\pi}{6} = -\frac{\sqrt{3}}{2}.$$

(c)
Reference number: $r = \frac{\pi}{3}$
Quadrant of $P(t)$ : IV, so $\sin t < 0$ and $\cos t > 0$
So

$$\sin\frac{5\pi}{3} = -\sin\frac{\pi}{3} = -\frac{\sqrt{3}}{2}, \qquad \cos\frac{5\pi}{3} = \cos\frac{\pi}{3} = \frac{1}{2}.$$

(d)
Reference number: $r = \frac{\pi}{4}$
Quadrant of $P(t)$ : I, so $\sin t > 0$ and $\cos t > 0$
So

$$\sin\left(-\frac{7\pi}{4}\right) = \sin\frac{\pi}{4} = \frac{\sqrt{2}}{2}, \qquad \cos\left(-\frac{7\pi}{4}\right) = \cos\frac{\pi}{4} = \frac{\sqrt{2}}{2}.$$

∎

**Example 128** *(Exercise Set 3.2, Exercise 26) Find all values of $t$ in the interval $[0, 2\pi]$ that satisfy the equation $\sin 3t = -\frac{\sqrt{2}}{2}$.*

Solution: First note that if $0 \le t \le 2\pi$, then $0 \le 3t \le 6\pi$. So, we need to consider values of $3t$ between $0$ and $6\pi$. Since

$$\sin 3t = -\frac{\sqrt{2}}{2}, \text{ for } 3t = \frac{5\pi}{4}, \frac{7\pi}{4}, \frac{13\pi}{4}, \frac{15\pi}{4}, \frac{21\pi}{4}, \frac{23\pi}{4}$$

we have

$$t = \frac{5\pi}{12}, \frac{7\pi}{12}, \frac{13\pi}{12}, \frac{15\pi}{12}, \frac{21\pi}{12}, \frac{23\pi}{12}.$$

∎

## 3.2. THE SINE AND COSINE FUNCTIONS

**Example 129** *(Exercise Set 3.2, Exercise 33) If $\sin t = \frac{3}{5}$ and $P(t)$ is in quadrant II, find $\cos t$.*

Solution: Since the $\cos t$ and $\sin t$ are the $x$- and $y$-coordinates, respectively, of points on the unit circle, $x^2 + y^2 = 1$, they satisfy the Pythagorean Identity

$$(\sin t)^2 + (\cos t)^2 = 1.$$

So

$$\left(\frac{3}{5}\right)^2 + (\cos t)^2 = 1$$

$$(\cos t)^2 = 1 - \frac{9}{25} = \frac{16}{25}$$

$$\cos t = \pm\sqrt{\frac{16}{25}} = \pm\frac{4}{5}.$$

Since $P(t)$ is in quadrant II, $\cos t < 0$, so

$$\cos t = -\frac{4}{5}.$$

■

**Example 130** *(Exercise Set 3.2, Exercise 37) Find all $t$ in the interval $[0, 2\pi]$, satisfying $(\cos t)^2 + \cos t - 2 = 0$.*

Solution: If we set $x = \cos t$, then the equation is the same as

$$x^2 + x - 2 = 0$$
$$(x-1)(x+2) = 0$$
$$x = 1, \quad x = -2.$$

So solve

$$\cos t = 1, \quad \cos t = -2.$$

The equation $\cos t = -2$ has no solutions since $|\cos t| \leq 1$ (recall the same holds for $\sin t$, that is, $|\sin t| \leq 1$). The solutions are,

$$\cos t = 1 \Rightarrow$$
$$t = 0, 2\pi.$$

■

## 3.3 Graphs of the Sine and Cosine Functions

A function $f$ is *periodic* if there is a positive number $T$ such that $f(t+T) = f(t)$, for all $t$ in the domain of $f$. The smallest such $T$ is called the *period* of the function. Since the circumference of the unit circle is $2\pi$,

$$\sin(t + 2\pi) = \sin t \quad \text{and} \quad \cos(t + 2\pi) = \cos t,$$

and the period of the sine and cosine functions is $2\pi$.

Three important features of the graphs of

$$y = A\sin(Bx + C), \quad \text{and} \quad y = A\cos(Bx + C),$$

where $A \neq 0$ and $B > 0$ are:

(1) <u>Amplitude:</u> The *amplitude* is the height of the sine or cosine wave which equals $|A|$.

(2) <u>Period:</u> The period of the sine or cosine function is $\frac{2\pi}{B}$.

(3) <u>Phase Shift:</u> To determine the amount the sine or cosine is horizontally shifted first write the function in the form

$$y = A\sin B\left(x + \frac{C}{B}\right), \quad \text{or} \quad y = A\cos B\left(x + \frac{C}{B}\right).$$

The graph is horizontally shifted $\frac{C}{B}$ units to the left if $\frac{C}{B} > 0$, and horizontally shifted $\left|\frac{C}{B}\right|$ to the right if $\frac{C}{B} < 0$.

**Example 131** *Use the graph of the sine and cosine to sketch one period of the graph of the given function.*

(a) *(Exercise Set 3.3, Exercise 4)*    $y = 2\cos x - 1$
(b) *(Exercise Set 3.3, Exercise 5)*    $y = 2\sin \frac{x}{2}$
(c) *(Exercise Set 3.3, Exercise 11)*    $y = -2 + 3\sin\left(3x - \frac{\pi}{2}\right)$

Solution:
(a)
<u>Amplitude:</u> 2, so multiply each y-coordinate of $y = \cos x$ by 2 which vertically stretches the graph.
<u>Period:</u> Same as $y = \cos x$, so $2\pi$.
<u>Phase Shift:</u> None.

To obtain the graph of $y = 2\cos x - 1$, vertically stretch $y = \cos x$ by a factor of 2, and then shift the resulting graph 1 unit vertically downward.

## 3.3. GRAPHS OF THE SINE AND COSINE FUNCTIONS

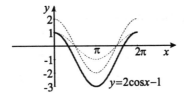
$y=2\cos x-1$

(b)
Amplitude: 2
Period:
$$\frac{2\pi}{B} = \frac{2\pi}{1/2} = 4\pi$$

The graph is horizontally stretched and completes one entire wave on the interval $[0, 4\pi]$.

Phase Shift: None.

It might be helpful to plot the points where the graph crosses the $x$-axis and the points where the wave is at its highest and lowest point.

$x$-intercepts :

$$2\sin\frac{x}{2} = 0$$
$$\sin\frac{x}{2} = 0$$
$$\frac{x}{2} = 0, \pi, 2\pi$$
$$x = 0, 2\pi, 4\pi$$

High and Low Points:

$$2\sin\frac{x}{2} = 2, \text{ or } 2\sin\frac{x}{2} = -2$$
$$\sin\frac{x}{2} = 1, \text{ or } \sin\frac{x}{2} = -1$$
$$\frac{x}{2} = \frac{\pi}{2}, \text{ or } \frac{x}{2} = \frac{3\pi}{2}$$
$$x = \pi, 3\pi$$

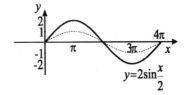
$y=2\sin\frac{x}{2}$

(c) First rewrite the function as

$$y = -2 + 3\sin\left(3x - \frac{\pi}{2}\right)$$
$$= -2 + 3\sin 3\left(x - \frac{\pi}{6}\right).$$

<u>Amplitude:</u> 3
<u>Period:</u>
$$\frac{2\pi}{B} = \frac{2\pi}{3}$$

The graph is horizontally compressed and completes one entire wave on an interval of length $\frac{2\pi}{3}$.

<u>Phase Shift:</u> $\frac{\pi}{6}$ units to the right.
<u>$x$-intercepts of $y = 3\sin\left(3x - \frac{\pi}{2}\right)$:</u>

$$3\sin\left(3x - \frac{\pi}{2}\right) = 0$$
$$\sin\left(3x - \frac{\pi}{2}\right) = 0$$
$$3x - \frac{\pi}{2} = 0, \pi, 2\pi$$
$$3x = \frac{\pi}{2}, \frac{3\pi}{2}, \frac{5\pi}{2}$$
$$x = \frac{\pi}{6}, \frac{\pi}{2}, \frac{5\pi}{6}$$

<u>High and Low points of $y = 3\sin\left(3x - \frac{\pi}{2}\right)$:</u>

$$3\sin\left(3x - \frac{\pi}{2}\right) = 3 \quad \text{or} \quad 3\sin\left(3x - \frac{\pi}{2}\right) = -3$$
$$\sin\left(3x - \frac{\pi}{2}\right) = 1 \quad \text{or} \quad \sin\left(3x - \frac{\pi}{2}\right) = -1$$
$$3x - \frac{\pi}{2} = \frac{\pi}{2} \quad \text{or} \quad 3x - \frac{\pi}{2} = \frac{3\pi}{2}$$
$$x = \frac{\pi}{3}, \frac{2\pi}{3}$$

Notice that $\frac{5\pi}{6} - \frac{\pi}{6} = \frac{4\pi}{6} = \frac{2\pi}{3}$, the period of the function. Also, $\frac{\pi}{3}$ is midway between $\frac{\pi}{6}$ and $\frac{\pi}{2}$ and $\frac{2\pi}{3}$ is midway between $\frac{\pi}{2}$ and $\frac{5\pi}{6}$.

To obtain the graph, start with the graph of $y = \sin x$, vertically stretch it by a factor of 3. Then horizontally compress the resulting graph so one

## 3.3. GRAPHS OF THE SINE AND COSINE FUNCTIONS

complete wave occurs on the interval $\left[0, \frac{2\pi}{3}\right]$. Shift this graph $\frac{\pi}{6}$ units to the right, and because of the $-2$; downward 2 units.

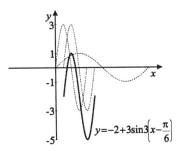

**Example 132** *(Exercise Set 3.3, Exercise 18) Find a sine or cosine function whose graph matches the curve.*

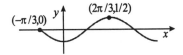

Solution: The curve has the appearance of a sine curve of the form $y = A \sin B \left(x + \frac{C}{B}\right)$, which has also been reflected about the $x$-axis.

<u>Amplitude:</u> The height of the wave is $\frac{1}{2}$, so $|A| = \frac{1}{2}$.

<u>Period:</u> One complete wave occurs on the interval $\left[-\frac{\pi}{3}, \pi\right]$, so the period is $\pi - \left(\frac{\pi}{3}\right) = \frac{4\pi}{3}$ and

$$\frac{2\pi}{B} = \frac{4\pi}{3}$$
$$B = \frac{3}{2}.$$

<u>Phase Shift:</u> The curve is shifted $\frac{\pi}{3}$ units to the left.
The equation of the curve is

$$y = -\frac{1}{2} \sin \frac{3}{2} \left(x + \frac{\pi}{3}\right).$$

The minus sign causes the reflection about the $x$-axis.

**Example 133** *(Exercise Set 3.3, Exercise 19) Determine an appropriate viewing rectangle for the $f(x) = \cos(100x)$ and use it to sketch the graph.*

Solution: Since there is no phase shift, the period of the function gives a good viewing rectangle, so compute

$$\frac{2\pi}{100} = \frac{\pi}{50} \approx 0.06.$$

One complete wave of the function occurs on the interval $\left[0, \frac{\pi}{50}\right]$ and an appropriate viewing rectangle is $\left[0, \frac{\pi}{50}\right] \times [-1, 1]$.

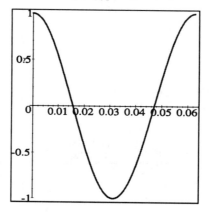

■

**Example 134** *(Exercise Set 3.3, Exercise 26) Use a graphing device to approximate the solutions to the equation $\sin x + x = x^3$.*

Solution: To approximate the solution use a graphing device to sketch the two curves $y = \sin x + x$ and $y = x^3$ and approximate the points of intersection.

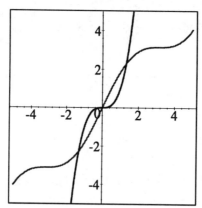

The points of intersection are

$$x = 0, \quad x \approx -1.32, \quad \text{and} \quad x \approx 1.32.$$

■

**Example 135** *(Exercise Set 3.3, Exercise 28)* Write each of the following as the composition of two functions $h(x) = f(g(x))$.
(a) $h(x) = \sqrt{\sin x}$   (b) $h(x) = 3\cos(4x - 2)$

Solution:
(a) The inside operation takes the sine of the number $x$ and the outside operation takes the square root. So define

$$f(x) = \sqrt{x} \quad \text{and} \quad g(x) = \sin x$$

and

$$f(g(x)) = f(\sin x) = \sqrt{\sin x}.$$

(b) The inside operation is the argument of the cosine function. Define

$$f(x) = 3\cos x \quad \text{and} \quad g(x) = 4x - 2$$

so

$$f(g(x)) = f(4x - 2) = 3\cos(4x - 2).$$

■

## 3.4 Other Trigonometric Functions

There are four additional trigonometric functions defined in terms of the sine and cosine.

$$\tan x = \frac{\sin x}{\cos x} \qquad \cot x = \frac{\cos x}{\sin x}$$

$$\sec x = \frac{1}{\cos x} \qquad \csc x = \frac{1}{\sin x}$$

The tangent and secant functions are defined whenever $x \neq \frac{\pi}{2} + n\pi$ for an integer $n$. The cotangent and cosecant are defined whenever $x \neq n\pi$ for an integer $n$.

**Example 136** *(Exercise Set 3.4, Exercise 8)* Find the values of all the trigonometric functions given that $\cos t = -\frac{1}{2}$ and $\pi \leq t \leq 2\pi$.

Solution: First find $\sin t$ using the Pythagorean Identity. That is,

$$(\cos t)^2 + (\sin t)^2 = 1$$
$$\left(-\frac{1}{2}\right)^2 + (\sin t)^2 = 1$$
$$(\sin t)^2 = 1 - \frac{1}{4} = \frac{3}{4}$$
$$\sin t = \pm\sqrt{\frac{3}{4}} = \pm\frac{\sqrt{3}}{\sqrt{4}}$$
$$\sin t = \pm\frac{\sqrt{3}}{2}.$$

Now apply the fact that $\pi \leq t \leq 2\pi$, so $t$ is in either quadrant III or IV. But $\sin t < 0$ in both quadrants. Note that $t$ is in quadrant III since $\cos t < 0$ in quadrant III and $\cos t > 0$ in quadrant IV. Thus

$$\cos t = -\frac{1}{2}, \quad \sin t = -\frac{\sqrt{3}}{2}$$

$$\tan t = \frac{-\frac{\sqrt{3}}{2}}{-\frac{1}{2}} = \sqrt{3}, \quad \cot t = \frac{-\frac{1}{2}}{-\frac{\sqrt{3}}{2}} = \frac{1}{2} \cdot \frac{2}{\sqrt{3}} = \frac{1}{\sqrt{3}} = \frac{1}{\sqrt{3}} \cdot \frac{\sqrt{3}}{\sqrt{3}} = \frac{\sqrt{3}}{3}$$

$$\sec t = \frac{1}{-\frac{1}{2}} = -2, \quad \csc t = \frac{1}{-\frac{\sqrt{3}}{2}} = -\frac{2}{\sqrt{3}} = -\frac{2\sqrt{3}}{3}.$$

■

### 3.4.1 Other Graphs

|  | Period | $x$-intercepts | Vertical asymptotes |
|---|---|---|---|
| $y = \tan x$ | $\pi$ | $x = \pm k\pi, k = 0, 1...$ | $x = \pm(2k-1)\frac{\pi}{2}, k = 1, 2...$ |
| $y = \cot x$ | $\pi$ | $x = \pm(2k-1)\frac{\pi}{2}, k = 1, 2...$ | $x = \pm k\pi, k = 0, 1...$ |
| $y = \sec x$ | $2\pi$ | none | $x = \pm(2k-1)\frac{\pi}{2}, k = 1, 2...$ |
| $y = \csc x$ | $2\pi$ | none | $x = \pm k\pi, k = 0, 1...$ |

## 3.4. OTHER TRIGONOMETRIC FUNCTIONS

**Example 137** *Sketch one period of the given curve.*
(a) $y = 2\tan\left(x - \frac{\pi}{2}\right)$
(b) $y = \sec(3x)$
(c) $y = \cot(2x + \pi)$

Solution:
(a)
Period: $\pi$, the same as the period of tangent.
Phase shift: $\frac{\pi}{2}$ units to the right.
The strategy is to use the graph of $y = \tan x$, vertically stretch it by a factor of 2, and then shift the resulting graph to the right $\frac{\pi}{2}$ units.
x-intercept:

$$2\tan\left(x - \frac{\pi}{2}\right) = 0$$

$$\tan\left(x - \frac{\pi}{2}\right) = \frac{\sin\left(x - \frac{\pi}{2}\right)}{\cos\left(x - \frac{\pi}{2}\right)} = 0$$

$$\sin\left(x - \frac{\pi}{2}\right) = 0$$

$$x - \frac{\pi}{2} = 0$$

$$x = \frac{\pi}{2}$$

This is exactly what is expected since $y = \tan x$ crosses the x-axis at $(0,0)$, and the graph of $y = 2\tan\left(x - \frac{\pi}{2}\right)$ is shifted $\frac{\pi}{2}$ units to the right. Here it crosses the x-axis at $\left(\frac{\pi}{2}, 0\right)$.

Asymptotes: $y = \tan x$ has asymptotes $x = -\frac{\pi}{2}$ and $x = \frac{\pi}{2}$, so $y = 2\tan\left(x - \frac{\pi}{2}\right)$ has asymptotes $x = -\frac{\pi}{2} + \frac{\pi}{2} = 0$ and $x = \frac{\pi}{2} + \frac{\pi}{2} = \pi$.

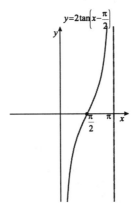

(b) One way to quickly plot $y = \sec 3x$ is to first plot $y = \cos 3x$. The graph of $y = \sec 3x$ is then a series of U-shaped curves that sit above and below the high points and low points, respectively, of the cosine wave.

Period: $\frac{2\pi}{3}$

Phase shift: none

Vertical asymptotes: Where $y = \cos 3x = 0$.

$$\begin{aligned} \cos 3x &= 0 \\ 3x &= \pm\frac{\pi}{2}, \pm\frac{3\pi}{2}, \pm\frac{5\pi}{2}, \ldots \\ x &= \pm\frac{\pi}{6}, \pm\frac{\pi}{2}, \pm\frac{5\pi}{6} \ldots \end{aligned}$$

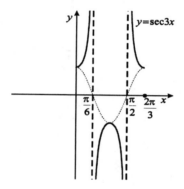

(c) Rewrite the function in form

$$y = \cot(2x + \pi) = \cot 2\left(x + \frac{\pi}{2}\right).$$

Period: $\frac{\pi}{2}$

Phase shift: $\frac{\pi}{2}$ units to the left.

One period of $y = \cot x$ occurs on the interval $(0, \pi)$, has vertical asymptotes at $x = 0$ and $x = \pi$, and crosses the $x$-axis at $x = \frac{\pi}{2}$. So $y = \cot 2x$, has period $\frac{\pi}{2}$, vertical asymptotes $x = 0$ and $x = \frac{\pi}{2}$, and crosses the $x$-axis at $x = \frac{\pi}{4}$. Now shift this graph $\frac{\pi}{2}$ units to the left.

Vertical asymptotes of $y = \cot 2\left(x + \frac{\pi}{2}\right)$ : $x = -\frac{\pi}{2}$ and $x = 0$.

$x$-intercept of $y = \cot 2\left(x + \frac{\pi}{2}\right)$ : $\left(-\frac{\pi}{4}, 0\right)$

## 3.4. OTHER TRIGONOMETRIC FUNCTIONS

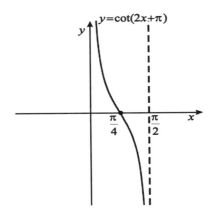

∎

**Example 138** *(Exercise Set 3.4, Exercise 25)* Find all values of $t$ in the interval $[0, 2\pi]$ that satisfy the equation $|\tan t| = 1$.

Solution:
$$|\tan t| = 1$$
$$\left|\frac{\sin t}{\cos t}\right| = 1$$
$$|\sin t| = |\cos t|$$

In the interval $[0, 2\pi]$ the sine and cosine have the same magnitude, equaling $\frac{\sqrt{2}}{2}$, at $\frac{\pi}{4}, \frac{3\pi}{4}, \frac{5\pi}{4}$, and $\frac{7\pi}{4}$. So

$$|\sin t| = |\cos t| = \frac{\sqrt{2}}{2}, \quad \text{for } t = \frac{\pi}{4}, \frac{3\pi}{4}, \frac{5\pi}{4}, \frac{7\pi}{4}.$$

∎

**Example 139** *(Exercise Set 3.4, Exercise 27)* Find all values of $t$ in the interval $[0, 2\pi]$ that satisfy the equation $2\sin 2t - \sqrt{2}\tan 2t = 0$.

Solution: We have the following sequence of equations.
$$2\sin 2t - \sqrt{2}\tan 2t = 0$$
$$2\sin 2t - \sqrt{2}\frac{\sin 2t}{\cos 2t} = 0$$
$$\frac{2\sin 2t \cos 2t - \sqrt{2}\sin 2t}{\cos 2t} = 0$$
$$2\sin 2t \cos 2t - \sqrt{2}\sin 2t = 0$$
$$\sin 2t(2\cos 2t - \sqrt{2}) = 0$$

Solutions to this equation occur when

$$\sin 2t = 0, \text{ or } \cos 2t = \frac{\sqrt{2}}{2}$$
$$2t = 0, \pi, 2\pi, \text{ or } 2t = \frac{\pi}{4}, \frac{7\pi}{4}$$

that is,

$$t = 0, \frac{\pi}{2}, \pi, \frac{\pi}{8}, \frac{7\pi}{8}$$

∎

**Example 140** *(Exercise Set 3.4, Exercise 35) Determine the values of the trigonometric functions of $t$ if $P(t)$ lies in the fourth quadrant and on the line $y = -2x$.*

Solution: To find the sine and cosine we need to determine the $x$- and $y$-coordinates of the point $P(t)$. Since the point lies on both the line and the unit circle, it must satisfy the two equations

$$y = -2x \quad \text{and} \quad x^2 + y^2 = 1.$$

Solving for $y$ in the first equation and substituting into the second for $y$ gives

$$x^2 + (-2x)^2 = 1$$
$$5x^2 = 1$$
$$x^2 = \frac{1}{5}$$
$$x = \pm\sqrt{\frac{1}{5}} = \pm\frac{\sqrt{5}}{5}.$$

Since $P(t)$ is in the fourth quadrant $\cos t > 0$ and the $x$-coordinate of $P(t)$ is greater than 0. So

$$x = \frac{\sqrt{5}}{5} \quad \text{and} \quad y = -\frac{2\sqrt{5}}{5}.$$

## 3.5. TRIGONOMETRIC IDENTITIES

The values of the trigonometric functions are

$$\sin t = -\frac{2\sqrt{5}}{5}, \quad \cos t = \frac{\sqrt{5}}{5}$$

$$\tan t = -2, \quad \cot t = -\frac{1}{2}$$

$$\sec t = \frac{5}{\sqrt{5}} = \sqrt{5}, \quad \csc t = -\frac{5}{2\sqrt{5}} = -\frac{\sqrt{5}}{2}.$$

∎

## 3.5 Trigonometric Identities

An *identity* is an equation that is true for every value of the variable. Trigonometric identities are equations that are true for all values of the variable for which both sides are defined.

### 3.5.1 Fundamental Identities

$$\tan x = \frac{\sin x}{\cos x} \qquad \cot x = \frac{\cos x}{\sin x}$$

$$\sec x = \frac{1}{\cos x} \qquad \csc x = \frac{1}{\sin x}$$

$$(\sin x)^2 + (\cos x)^2 = 1 \quad 1 + (\tan x)^2 = (\sec x)^2 \quad 1 + (\cot x)^2 = (\csc x)^2$$

$$\sin(-x) = -\sin x \qquad \cos(-x) = \cos x \qquad \tan(-x) = -\tan x$$

**Example 141** *Verify the identities.*
(a) $(1 - (\cos x)^2)(\sec x)^2 = (\tan x)^2$
(b) $\tan x + \cot x = \sec x \csc x$
(c) $\dfrac{\cos x}{1 - \tan x} + \dfrac{\sin x}{1 - \cot x} = \sin x + \cos x$
(d) (Exercise Set 3.5, Exercise 40) $(\tan x)^2 - (\sin x)^2 = (\tan x)^2 (\sin x)^2$

Solution:
(a) One method for verifying identities is to start with one side, replace any trigonometric functions with sines and cosines and then try to reduce the resulting expression to that on the other side of the identity. Be on the look out for the Fundamental Identities, especially the Pythagorean Identities.

In this example start with the left side.

$$(1 - (\cos x)^2)(\sec x)^2 = (1 - (\cos x)^2)\frac{1}{(\cos x)^2}$$
$$= (\sin x)^2 \frac{1}{(\cos x)^2}$$
$$= \frac{(\sin x)^2}{(\cos x)^2}$$
$$= (\tan x)^2$$

We used one of the equivalent forms of the Pythagorean Identity

$$(\sin x)^2 + (\cos x)^2 = 1,$$

in going from the first line to the second. That is,

$$(\sin x)^2 + (\cos x)^2 = 1 \Rightarrow$$
$$(\sin x)^2 = 1 - (\cos x)^2 \quad \text{and} \quad (\cos x)^2 = 1 - (\sin x)^2.$$

(b) First replace the left side with sines and cosines.

$$\tan x + \cot x = \frac{\sin x}{\cos x} + \frac{\cos x}{\sin x}$$

Now take the common denominator of the two fractions. Recall for any algebraic expressions

$$\frac{a}{b} + \frac{c}{d} = \frac{ad + cb}{bd}.$$

So

$$\tan x + \cot x = \frac{\sin x}{\cos x} + \frac{\cos x}{\sin x}$$
$$= \frac{\sin x \sin x + \cos x \cos x}{\cos x \sin x}$$
$$= \frac{(\sin x)^2 + (\cos x)^2}{\cos x \sin x}$$
$$= \frac{1}{\cos x \sin x}$$
$$= \frac{1}{\cos x} \cdot \frac{1}{\sin x}$$
$$= \sec x \csc x.$$

## 3.5. TRIGONOMETRIC IDENTITIES

(c) Changing to sines and cosines gives

$$\frac{\cos x}{1-\tan x} + \frac{\sin x}{1-\cot x} = \frac{\cos x}{1-\frac{\sin x}{\cos x}} + \frac{\sin x}{1-\frac{\cos x}{\sin x}}$$

$$= \frac{\cos x}{\frac{\cos x - \sin x}{\cos x}} + \frac{\sin x}{\frac{\sin x - \cos x}{\sin x}}$$

$$= \cos x \cdot \frac{\cos x}{\cos x - \sin x} + \sin x \cdot \frac{\sin x}{\sin x - \cos x}$$

$$= \frac{(\cos x)^2}{\cos x - \sin x} + \frac{(\sin x)^2}{\sin x - \cos x}$$

If the last two fractions had the same denominator they could be added. Since

$$\sin x - \cos x = -(\cos x - \sin x)$$

we have

$$\frac{\cos x}{1-\tan x} + \frac{\sin x}{1-\cot x} = \frac{(\cos x)^2}{\cos x - \sin x} + \frac{(\sin x)^2}{\sin x - \cos x}$$

$$= \frac{(\cos x)^2}{\cos x - \sin x} - \frac{(\sin x)^2}{\cos x - \sin x}$$

$$= \frac{(\cos x)^2 - (\sin x)^2}{\cos x - \sin x}$$

$$= \frac{(\cos x - \sin x)(\cos x + \sin x)}{\cos x - \sin x}$$

$$= \cos x + \sin x.$$

In the next to the last step we used the factoring formula

$$a^2 - b^2 = (a-b)(a+b).$$

(d) We again change to sines and cosines

$$(\tan x)^2 - (\sin x)^2 = \frac{(\sin x)^2}{(\cos x)^2} - (\sin x)^2$$

$$= \frac{(\sin x)^2 - (\sin x)^2(\cos x)^2}{(\cos x)^2}$$

$$= \frac{(\sin x)^2(1-(\cos x)^2)}{(\cos x)^2}$$

$$\begin{aligned}
&= \frac{(\sin x)^2}{(\cos x)^2} \cdot (1 - (\cos x)^2) \\
&= \left(\frac{\sin x}{\cos x}\right)^2 \cdot (1 - (\cos x)^2) \\
&= (\tan x)^2 (\sin x)^2
\end{aligned}$$

■

The following substitutions are frequently useful in calculus to simplify expressions involving radicals.

**Example 142** *Make the indicated trigonometric substitution and simplify the expression.*

(a) $\sqrt{1 - x^2}$; $x = \sin t$, for $-\frac{\pi}{2} \leq t \leq \frac{\pi}{2}$

(b) $\dfrac{\sqrt{x^2 - 1}}{x}$; $x = \sec t$, for $0 < t < \frac{\pi}{2}$

(c) $\dfrac{x}{(1 - x^2)^{3/2}}$; $x = \sin t$, for $-\frac{\pi}{2} < t < \frac{\pi}{2}$

Solution:
(a)

$$\begin{aligned}
\sqrt{1 - x^2} &= \sqrt{1 - (\sin t)^2} \\
&= \sqrt{(\cos t)^2} \\
&= |\cos t| \\
&= \cos t
\end{aligned}$$

The absolute value can be dropped since for $-\frac{\pi}{2} \leq t \leq \frac{\pi}{2}$ we have $\cos t \geq 0$.

(b)

$$\begin{aligned}
\frac{\sqrt{x^2 - 1}}{x} &= \frac{\sqrt{(\sec t)^2 - 1}}{\sec t} \\
&= \frac{\sqrt{(\tan t)^2}}{\sec t} \\
&= \frac{\tan t}{\sec t} \\
&= \frac{\frac{\sin t}{\cos t}}{\frac{1}{\cos t}}
\end{aligned}$$

## 3.5. TRIGONOMETRIC IDENTITIES

$$= \frac{\sin t}{\cos t} \cdot \frac{\cos t}{1}$$
$$= \sin t$$

(c)

$$\frac{x}{(1-x^2)^{3/2}} = \frac{\sin t}{(1-(\sin t)^2)^{3/2}}$$
$$= \frac{\sin t}{((\cos t)^2)^{3/2}}$$
$$= \frac{\sin t}{(\cos t)^3}$$
$$= \frac{\sin t}{\cos t} \cdot \frac{1}{(\cos t)^2}$$
$$= \tan t (\sec t)^2$$

■

### 3.5.2 Addition and Subtraction Identities

The fundamental addition and subtraction identities are

$$\sin(x \pm y) = \sin x \cos y \pm \cos x \sin y$$
$$\cos(x \pm y) = \cos x \cos y \mp \sin x \sin y.$$

These can be used to show that

$$\tan(x \pm y) = \frac{\tan x \pm \tan y}{1 \mp \tan x \tan y},$$

as well as identities involving the other trigonometric functions.

**Example 143** *Determine the exact value of the trigonometric function.*
(a) *(Exercise Set 3.5, Exercise 2)* $\cos\left(\frac{5\pi}{6} + \frac{\pi}{4}\right)$
(b) *(Exercise Set 3.5, Exercise 3)* $\sin\left(\frac{7\pi}{12}\right)$

Solution:

(a) Applying the addition formula for cosine

$$\cos\left(\frac{5\pi}{6}+\frac{\pi}{4}\right) = \cos\left(\frac{5\pi}{6}\right)\cos\left(\frac{\pi}{4}\right) - \sin\left(\frac{5\pi}{6}\right)\sin\left(\frac{\pi}{4}\right)$$
$$= -\frac{\sqrt{3}}{2}\cdot\frac{\sqrt{2}}{2} - \frac{1}{2}\cdot\frac{\sqrt{2}}{2}$$
$$= -\frac{\sqrt{2}}{4}\left(\sqrt{3}+1\right).$$

(b) First express $\frac{7\pi}{12}$ as the sum of two values for which we know the exact values of the sine and cosine:

$$\frac{7\pi}{12} = \frac{\pi}{3}+\frac{\pi}{4}.$$

Then

$$\sin\left(\frac{7\pi}{12}\right) = \sin\left(\frac{\pi}{3}+\frac{\pi}{4}\right)$$
$$= \sin\left(\frac{\pi}{3}\right)\cos\left(\frac{\pi}{4}\right) + \cos\left(\frac{\pi}{3}\right)\sin\left(\frac{\pi}{4}\right)$$
$$= \frac{\sqrt{3}}{2}\cdot\frac{\sqrt{2}}{2} + \frac{1}{2}\cdot\frac{\sqrt{2}}{2}$$
$$= \frac{\sqrt{2}}{4}\left(\sqrt{3}+1\right).$$

∎

**Example 144** *Express $\sin 3x$ in terms of $\sin x$ and $\cos x$.*

Solution: First write $3x = 2x + x$ and apply the addition formula for the sine function.

$$\sin 3x = \sin(2x+x)$$
$$= \sin 2x \cos x + \cos 2x \sin x$$

We now apply the addition formula a second time to $\sin 2x = \sin(x+x)$ and $\cos 2x = \cos(x+x)$. Then

$$\sin 3x = \sin 2x \cos x + \cos 2x \sin x$$
$$= (\sin x \cos x + \cos x \sin x)\cos x + (\cos x \cos x - \sin x \sin x)\sin x$$
$$= 2\sin x(\cos x)^2 + \sin x(\cos x)^2 - (\sin x)^3$$
$$= 3\sin x(\cos x)^2 - (\sin x)^3.$$

## 3.5. TRIGONOMETRIC IDENTITIES

■

**Example 145** *Use the addition and subtraction formulas to verify the given identity.*

*(a) (Exercise Set 3.5, Exercise 15)* $\sin\left(t + \frac{\pi}{2}\right) = \cos t$
*(b) (Exercise Set 3.5, Exercise 18)* $\cos(\pi - t) = -\cos t$

Solution:
(a) We have

$$\begin{aligned}\sin\left(t + \frac{\pi}{2}\right) &= \sin t \cos\left(\frac{\pi}{2}\right) + \cos t \sin\left(\frac{\pi}{2}\right) \\ &= (\sin t) \cdot (0) + (\cos t) \cdot (1) \\ &= \cos t.\end{aligned}$$

(b) We have

$$\begin{aligned}\cos(\pi - t) &= \cos\pi \cos t + \sin\pi \sin t \\ &= (-1) \cdot \cos t + (0) \cdot \sin t \\ &= -\cos t.\end{aligned}$$

■

### 3.5.3 Double Angle Formulas

The basic double-angle formulas are

$$\sin 2x = 2\sin x \cos x$$
$$\cos 2x = (\cos x)^2 - (\sin x)^2 = 2(\cos x)^2 - 1 = 1 - 2(\sin x)^2.$$

These can be used to derive

$$\tan 2x = \frac{2\tan x}{1 - (\tan x)^2},$$

as well as identities involving the other trigonometric functions.

**Example 146** *If* $\cos t = \frac{3}{5}$, *where* $0 < t < \frac{\pi}{2}$, *find* $\cos 2t$, $\sin 2t$, *and* $\tan 2t$.

Solution: First find $\sin t$, which is needed in the double-angle formulas. Since $t$ is in the first quadrant, $\sin t > 0$. So

$$\begin{aligned} \sin t &= \sqrt{1 - (\cos t)^2} \\ &= \sqrt{1 - \left(\frac{3}{5}\right)^2} \\ &= \sqrt{\frac{16}{25}} \\ &= \frac{\sqrt{16}}{\sqrt{25}} \\ &= \frac{4}{5}. \end{aligned}$$

Then

$$\begin{aligned} \cos 2t &= 2\left(\frac{3}{5}\right)^2 - 1 = \frac{18}{25} - 1 = -\frac{7}{25} \\ \sin 2t &= 2\left(\frac{4}{5}\right)\left(\frac{3}{5}\right) = \frac{24}{25} \\ \tan 2t &= \frac{\sin 2t}{\cos 2t} = \frac{\frac{24}{25}}{-\frac{7}{25}} = -\frac{24}{25} \cdot \frac{25}{7} = -\frac{24}{7}. \end{aligned}$$

■

**Example 147** *(Exercise Set 3.5, Exercise 38) Verify the identity* $(\sin x + \cos x)^2 = 1 + \sin 2x$.

Solution: We have

$$\begin{aligned} (\sin x + \cos x)^2 &= (\sin x)^2 + 2\sin x \cos x + (\cos x)^2 \\ &= (\sin x)^2 + (\cos x)^2 + 2\sin x \cos x \\ &= 1 + 2\sin x \cos x \\ &= 1 + \sin 2x. \end{aligned}$$

■

## 3.5. TRIGONOMETRIC IDENTITIES

### 3.5.4 Half Angle Formulas

The basic half-angle formulas can be written either as

$$(\sin x)^2 = \frac{1 - \cos 2x}{2} \quad \text{and} \quad (\cos x)^2 = \frac{1 + \cos 2x}{2}$$

or as

$$\sin\left(\frac{x}{2}\right) = \sqrt{\frac{1 - \cos x}{2}} \quad \text{and} \quad \cos\left(\frac{x}{2}\right) = \sqrt{\frac{1 + \cos x}{2}}.$$

From these we can derive

$$(\tan x)^2 = \frac{1 - \cos 2x}{1 + \cos 2x} \quad \text{and} \quad \tan\left(\frac{x}{2}\right) = \frac{1 - \cos x}{\sin x} = \frac{\sin x}{1 + \cos x},$$

as well as identities involving the other trigonometric functions.

**Example 148** *(Exercise Set 3.5, Exercise 8) Find the exact value of* $\cos\left(\frac{\pi}{8}\right)$.

Solution: If $x = \frac{\pi}{4}$ in the formula for $\cos\left(\frac{x}{2}\right)$, then $\frac{x}{2} = \frac{\pi}{8}$ and

$$\begin{aligned}
\cos\left(\frac{\pi}{8}\right) &= \sqrt{\frac{1 + \cos\left(\frac{\pi}{4}\right)}{2}} \\
&= \sqrt{\frac{1 + \frac{\sqrt{2}}{2}}{2}} = \sqrt{\frac{2 + \sqrt{2}}{4}} \\
&= \frac{\sqrt{2 + \sqrt{2}}}{2}.
\end{aligned}$$

■

The technique in the next example is used to reduce powers of trigonometric functions to forms that are easier to work with.

**Example 149** *(Exercise Set 3.5, Exercise 21) Rewrite the expression* $(\cos x)^4$ *so that it involves the sum or difference of only constants and sine and cosine functions to the first power.*

Solution: Use the half angle-formula for the cosine to write

$$\begin{aligned}(\cos x)^4 &= ((\cos x)^2)^2 \\ &= \left(\frac{1+\cos 2x}{2}\right)^2 \\ &= \frac{1}{4}\left(1 + 2\cos 2x + (\cos 2x)^2\right).\end{aligned}$$

Now use the half-angle formula again on $(\cos 2x)^2$, so

$$(\cos 2x)^2 = \frac{1+\cos 4x}{2}$$

and

$$\begin{aligned}(\cos x)^4 &= \frac{1}{4}\left(1 + 2\cos 2x + (\cos 2x)^2\right) \\ &= \frac{1}{4}\left(1 + 2\cos 2x + \frac{1+\cos 4x}{2}\right) \\ &= \frac{1}{4} + \frac{\cos 2x}{2} + \frac{1+\cos 4x}{8} \\ &= \frac{3}{8} + \frac{\cos 2x}{2} + \frac{\cos 4x}{8}.\end{aligned}$$

∎

### 3.5.5 Product-to-Sum and Sum-to-Product Identities

The basic product-to-sum identities are

$$\sin x \cos y = \tfrac{1}{2}[\sin(x+y) + \sin(x-y)], \quad \cos x \sin y = \tfrac{1}{2}[\sin(x+y) - \sin(x-y)]$$
$$\cos x \cos y = \tfrac{1}{2}[\cos(x+y) + \cos(x-y)], \quad \sin x \sin y = \tfrac{1}{2}[\cos(x-y) - \cos(x+y)].$$

and the basic sum-to-product identities are

$$\sin x + \sin y = 2\sin\frac{x+y}{2}\cos\frac{x-y}{2}, \quad \sin x - \sin y = 2\cos\frac{x+y}{2}\sin\frac{x-y}{2}$$
$$\cos x + \cos y = 2\cos\frac{x+y}{2}\cos\frac{x-y}{2}, \quad \cos x - \cos y = -2\sin\frac{x+y}{2}\sin\frac{x-y}{2}.$$

## 3.5. TRIGONOMETRIC IDENTITIES

**Example 150** *(Exercise Set 3.5, Exercise 24)* *Write the product $\cos 5t \sin 8t$ as a sum or difference.*

Solution: We have

$$\cos 5t \sin 8t = \frac{1}{2}[\sin(5t + 8t) - \sin(5t - 8t)]$$
$$= \frac{1}{2}[\sin(13t) - \sin(-3t)]$$
$$= \frac{1}{2}\sin(13t) + \frac{1}{2}\sin(3t)$$

Recall $\sin(-x) = -\sin x$.
■

**Example 151** *(Exercise Set 3.5, Exercise 29)* *Rewrite $\cos 5t + \cos 2t$ as a product.*

Solution: We have

$$\cos 5t + \cos 2t = 2\cos\frac{5t+2t}{2}\cos\frac{5t-2t}{2}$$
$$= 2\cos\frac{7t}{2}\cos\frac{3t}{2}$$

■

### 3.5.6 Solving Trigonometric Equations

**Example 152** *(Exercise Set 3.5, Exercise 31)* *Find all values of $x$ in the interval $[0, 2\pi]$ that satisfy the equation $\sin 2x = \sin x$.*

Solution: Use the double-angle formula to rewrite $\sin 2x = 2\sin x \cos x$. Then

$$\sin 2x = \sin x$$
$$2\sin x \cos x = \sin x$$
$$2\sin x \cos x - \sin x = 0$$
$$\sin x (2\cos x - 1) = 0$$
$$\sin x = 0, \quad \cos x = \frac{1}{2}$$
$$x = 0, \pi, 2\pi \quad \text{or} \quad x = \frac{\pi}{3}, \frac{5\pi}{3}.$$

■

**Example 153** *Find all values of $x$ in the interval $[0, 2\pi]$ that satisfy the equation $\cos 2x \cos 3x = \sin 2x \sin 3x$.*

Solution: If we bring the expression on the right to the left side of the equation we recognize the new expression as fitting the sum formula for cosines. That is,

$$\begin{aligned}
\cos 2x \cos 3x - \sin 2x \sin 3x &= 0 \\
\cos(2x + 3x) &= 0 \\
\cos(5x) &= 0 \\
5x &= \frac{\pi}{2}, \frac{3\pi}{2}, \frac{5\pi}{2}, \frac{7\pi}{2}, \frac{9\pi}{2}, \frac{11\pi}{2}, \frac{13\pi}{2}, \frac{15\pi}{2}, \frac{17\pi}{2}, \frac{19\pi}{2} \\
x &= \frac{\pi}{10}, \frac{3\pi}{10}, \frac{7\pi}{10}, \frac{9\pi}{10}, \frac{11\pi}{10}, \frac{13\pi}{10}, \frac{17\pi}{10}, \frac{19\pi}{10}.
\end{aligned}$$

■

## 3.6 Right Triangle Trigonometry

### 3.6.1 Angle Measure: Radians and Degrees

An angle formed by rotating a ray counterclockwise from the initial position one complete revolution back to the initial position has, by definition, a measure of 360 degrees, written 360°. The radian measure of the same angle is equal to the circumference of the unit circle, which is $2\pi$ radians. Angles measured in the clockwise direction are negative.

<u>Conversion between Degrees and Radians</u>
$180° = \pi$ radians,  $1° = \frac{\pi}{180}$ radians,  $1$ radian $= \frac{180°}{\pi}$

**Example 154** *Convert from degree measure to radian measure.*
*(a) 45° (b) 300°*

Solution:

## 3.6. RIGHT TRIANGLE TRIGONOMETRY

(a) Since $1° = \frac{\pi}{180}$ radians,

$$45° = 45 \cdot \frac{\pi}{180}$$
$$= \frac{\pi}{4} \text{ radians.}$$

(b) Also,

$$300° = 300 \cdot \frac{\pi}{180}$$
$$= \frac{5\pi}{3} \text{ radians.}$$

■

**Example 155** *Convert from radian measure to degree measure.*
*(a) $\frac{5\pi}{4}$  (b) $-\frac{7\pi}{6}$*

Solution:
(a) Since 1 radian $= \frac{180°}{\pi}$,

$$\frac{5\pi}{4} \text{ radians} = \frac{5\pi}{4} \cdot \frac{180}{\pi}$$
$$= 225°.$$

(b) Similarly,

$$-\frac{7\pi}{6} \text{ radians} = -\frac{7\pi}{6} \cdot \frac{180}{\pi}$$
$$= -210°$$

■

### 3.6.2 Trigonometric Functions of an Angle in a Right Triangle

For an angle $\theta$ in the right triangle shown below we have the following values of the trigonometric functions.

$$\sin\theta = \frac{b}{c} = \frac{\text{opposite}}{\text{hypotenuse}} \qquad \cos\theta = \frac{a}{c} = \frac{\text{adjacent}}{\text{hypotenuse}}$$

$$\tan\theta = \frac{b}{a} = \frac{\text{opposite}}{\text{adjacent}} \qquad \cot\theta = \frac{a}{b} = \frac{\text{adjacent}}{\text{opposite}}$$

$$\sec\theta = \frac{c}{a} = \frac{\text{hypotenuse}}{\text{adjacent}} \qquad \csc\theta = \frac{c}{b} = \frac{\text{hypotenuse}}{\text{opposite}}$$

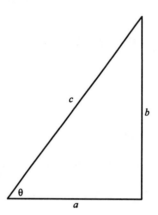

**Example 156** *Find the value of the six trigonometric functions of the angle $\theta$ shown in the following right triangle.*

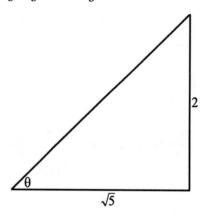

Solution: The length of the hypotenuse is needed and can be found using the Pythagorean Theorem. If $x$ denotes the length of the hypotenuse, then

$$\begin{aligned} x^2 &= \left(\sqrt{5}\right)^2 + 2^2 \\ &= 9 \\ x &= 3. \end{aligned}$$

## 3.6. RIGHT TRIANGLE TRIGONOMETRY

So

$$\sin\theta = \frac{2}{3}, \quad \cos\theta = \frac{\sqrt{5}}{3}$$

$$\tan\theta = \frac{2}{\sqrt{5}} = \frac{2\sqrt{5}}{5}, \quad \cot\theta = \frac{\sqrt{5}}{2}$$

$$\sec\theta = \frac{3}{\sqrt{5}} = \frac{3\sqrt{5}}{5}, \quad \csc\theta = \frac{3}{2}.$$

∎

**Example 157** *(Exercise Set 3.6, Exercise 19) Refer to the triangle in the figure and use the information to find any missing angles or sides.*

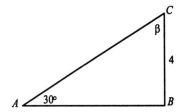

Solution: Since the sum of the angles of a triangle is 180°,

$$\beta = 180° - 90° - 30° = 60°.$$

To find the side $\overline{AB}$, use the tangent of the 30° angle. So

$$\tan 30° = \frac{4}{\overline{AB}}$$

$$\overline{AB} = \frac{4}{\tan 30°}$$
$$= \frac{4}{\sqrt{3}/3} = \frac{12}{\sqrt{3}} = 4\sqrt{3}.$$

Recall,

$$\begin{aligned} \tan 30° &= \tan \frac{\pi}{6} \\ &= \frac{\sin \frac{\pi}{6}}{\cos \frac{\pi}{6}} = \frac{1/2}{\sqrt{3}/2} \\ &= \frac{1}{\sqrt{3}} = \frac{1}{\sqrt{3}} \cdot \frac{\sqrt{3}}{\sqrt{3}} = \frac{\sqrt{3}}{3}. \end{aligned}$$

The remaining side $\overline{AC}$ can be found using the Pythagorean Theorem. That is,

$$\begin{aligned} \overline{AC}^2 &= \left(4\sqrt{3}\right)^2 + 4^2 \\ &= 48 + 16 \\ &= 64 \Rightarrow \\ \overline{AC} &= 8. \end{aligned}$$

■

### 3.6.3 Applications

**Example 158** *(Exercise Set 3.6, Exercise 26) A pipe line is to be constructed between points $A$ and $B$ on opposite sides of a river as shown in the figure. To determine the amount of pipe needed using a transit at point $A$, a line perpendicular to $AB$ is determined and point $C$ is marked approximately 300 feet from $A$. At point $C$ the transit is used to determine that the angle $\theta$ is $42°$. Determine the amount of pipe needed to connect $A$ and $B$.*

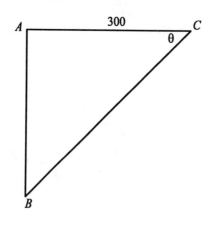

## 3.6. RIGHT TRIANGLE TRIGONOMETRY

Solution: The lengths $\overline{AB}$ and $\overline{AC} = 300$ are the sides opposite and adjacent, respectively, to the angle of 42° and so can be related using the tangent function. That is,

$$\tan 42° = \frac{\overline{AB}}{\overline{AC}} = \frac{\overline{AB}}{300} \Rightarrow$$
$$\overline{AB} = 300 \tan 42° \approx 270.12 \text{ feet.}$$

The amount of pipe needed to connect points $A$ and $B$ is approximately 270.12 feet. ∎

**Example 159** *(Exercise Set 3.6, Exercise 28)* An engineer is designing a drainage canal that has a trapezoidal cross section. The bottom and sides of the canal are each 10 feet long, and the side makes an angle $\theta$ with the horizontal. Find an expression for the cross sectional area of the channel in terms of the angle $\theta$. Use a graphing device to approximate the angle $\theta$ that will maximize the capacity of the canal.

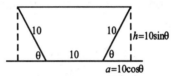

Solution: The area of a trapezoid is

$$\text{area of trapezoid} = \frac{1}{2}(\text{sum of the bases}) \times (\text{height}).$$

Let the cross sectional area of the canal be denoted $A$. From the figure

$$A = \frac{1}{2}[10 + (10 + 2a)]h$$
$$= \frac{1}{2}(20 + 2a)h$$
$$= (10 + a)h.$$

To express the area in terms of $\theta$ use trigonometric functions to express $a$ and $h$ in terms of $\theta$. So

$$\sin \theta = \frac{h}{10} \quad \text{and} \quad \cos \theta = \frac{a}{10}$$
$$h = 10 \sin \theta \quad \text{and} \quad a = 10 \cos \theta$$
$$A = (10 + 10 \cos \theta)(10 \sin \theta)$$
$$= 100(1 + \cos \theta) \sin \theta.$$

The capacity of the canal is the volume of water the canal can hold. If the canal is $S$ feet in length, then the volume of water, $V$, is

$$V = A \cdot S = 100S(1 + \cos\theta)\sin\theta \text{ ft}^3.$$

The capacity will be a maximum precisely for that value of $\theta$ that makes the function

$$f(\theta) = (1 + \cos\theta)\sin\theta$$

as large as possible. To approximate the value of $\theta$ we use a graphing device to plot the function and select the $\theta$, which gives the highest point on the curve.

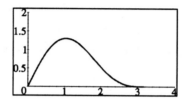

by refining the graph shown in the figure, we can show that the maximum value for $f(\theta)$ occurs when

$$\begin{aligned}\theta &\approx 1.047 \text{ radians} \\ &= 1.047\left(\frac{180}{\pi}\right) \text{ degrees} \\ &\approx 59.99°.\end{aligned}$$

The maximum capacity of the canal is approximately

$$\begin{aligned}V &\approx 100S(1 + \sin(1.047))\cos(1.047) \\ &\approx 93.3S \text{ cubic feet.}\end{aligned}$$

■

## 3.7 Inverse Trigonometric Functions

For a function to have an inverse the function must be one-to-one. The trigonometric functions are not one-to-one, which is easily seen by the horizontal line test. For each of the trigonometric functions there are horizontal lines that intersect the graph in many points, not just one. To define the inverse trigonometric functions the domains have to be restricted so the functions become one-to-one.

## 3.7. INVERSE TRIGONOMETRIC FUNCTIONS

### 3.7.1 Inverse Sine

The domain of the sine function is restricted to the interval $\left[-\frac{\pi}{2}, \frac{\pi}{2}\right]$, making it one-to-one. The inverse sine function is then defined by

$$y = \arcsin x \Leftrightarrow \sin y = x.$$

The inverse sine undoes the process of the restricted sine function.

An important relationship to remember about all functions and their inverses, which is very helpful when trying to construct graphs, is that

the domain of the function is the range of the inverse,

and

the range of the function is the domain of the inverse.

For the sine function

$$\begin{aligned}
\text{domain of arcsin} &= \text{range of sin} = [-1, 1] \\
\text{range of arcsin} &= \text{domain of sin} = -\frac{\pi}{2}, \frac{\pi}{2} \\
\arcsin(\sin x) &= x, \quad \text{for } x \text{ in the domain of sin} \\
\sin(\arcsin x) &= x, \quad \text{for } x \text{ in the domain of arcsin}.
\end{aligned}$$

**Example 160** *Find the exact value of the quantity.*
(a) $\arcsin\left(\frac{\sqrt{3}}{2}\right)$   (b) $\sin\left(\arcsin\left(\frac{1}{2}\right)\right)$
(c) $\arcsin\left(\sin \frac{5\pi}{4}\right)$   (d) $\cos\left(\arcsin\left(\frac{1}{2}\right)\right)$

Solution:
(a) The problem can be read as "find a number between $-\frac{\pi}{2}$ and $\frac{\pi}{2}$ whose sine is equal to $\frac{\sqrt{3}}{2}$." Since $\sin \frac{\pi}{3} = \frac{\sqrt{3}}{2}$,

$$\arcsin\left(\frac{\sqrt{3}}{2}\right) = \frac{\pi}{3}.$$

(b) The domain of arcsin is $[-1, 1]$, which contains $\frac{1}{2}$ so,

$$\sin\left(\arcsin\left(\frac{1}{2}\right)\right) = \frac{1}{2}.$$

(c) The answer to this part is *not* $\frac{5\pi}{4}$ since $\frac{5\pi}{4}$ is not in the restricted domain of the sine function. To solve this problem we first need to find $\sin\left(\frac{5\pi}{4}\right)$, and then find a number in $\left[-\frac{\pi}{2}, \frac{\pi}{2}\right]$ whose sine is the same as that of $\sin\left(\frac{5\pi}{4}\right)$. Since

$$\sin\frac{5\pi}{4} = -\frac{\sqrt{2}}{2}$$

$$\arcsin\left(\sin\frac{5\pi}{4}\right) = \arcsin\left(-\frac{\sqrt{2}}{2}\right) = -\frac{\pi}{4}.$$

Recall the sine is negative in quadrant IV.

(d)
$$\cos\left(\arcsin\left(\frac{1}{2}\right)\right) = \cos\left(\frac{\pi}{6}\right) = \frac{\sqrt{3}}{2}$$

■

## 3.7.2 Summary of the Inverse Trigonometric Functions

The table summarizes the definitions of all the inverse trigonometric functions.

| Definition | Domain | Range |
|---|---|---|
| $\arcsin x = y \Leftrightarrow \sin y = x$ | $[-1, 1]$ | $\left[-\frac{\pi}{2}, \frac{\pi}{2}\right]$ |
| $\arccos x = y \Leftrightarrow \cos y = x$ | $[-1, 1]$ | $[0, \pi]$ |
| $\arctan x = y \Leftrightarrow \tan y = x$ | $(-\infty, \infty)$ | $\left(-\frac{\pi}{2}, \frac{\pi}{2}\right)$ |
| $\text{arccot } x = y \Leftrightarrow \cot y = x$ | $(-\infty, \infty)$ | $(0, \pi)$ |
| $\text{arcsec } x = y \Leftrightarrow \sec y = x$ | $(-\infty, -1] \cup [1, \infty)$ | $\left[0, \frac{\pi}{2}\right) \cup \left(\frac{\pi}{2}, \pi\right]$ |
| $\text{arccsc } x = y \Leftrightarrow \csc y = x$ | $(-\infty, -1] \cup [1, \infty)$ | $\left(0, \frac{\pi}{2}\right] \cup \left(\pi, \frac{3\pi}{2}\right]$ |

**Example 161** *Find the exact value of each expression.*
(a) $\arccos\left(\cos\left(\frac{\pi}{6}\right)\right)$  (b) $\sin\left(\arctan\left(\sqrt{3}\right)\right)$  (c) $\arctan\left(\tan\left(\frac{5\pi}{6}\right)\right)$

Solution:
(a) The restricted domain of the cosine function is $[0, \pi]$, which contains $\frac{\pi}{6}$ so

$$\arccos\left(\cos\left(\frac{\pi}{6}\right)\right) = \frac{\pi}{6}.$$

## 3.7. INVERSE TRIGONOMETRIC FUNCTIONS

(b) The domain of arctan is all real numbers and since

$$\tan\left(\frac{\pi}{3}\right) = \frac{\sin\left(\frac{\pi}{3}\right)}{\cos\left(\frac{\pi}{3}\right)} = \frac{\frac{\sqrt{3}}{2}}{\frac{1}{2}} = \sqrt{3},$$

we have

$$\sin\left(\arctan\left(\sqrt{3}\right)\right) = \sin\left(\frac{\pi}{3}\right) = \frac{\sqrt{3}}{2}.$$

(c) Since the restricted domain of tan is $\left(-\frac{\pi}{2}, \frac{\pi}{2}\right)$, which does not contain $\frac{5\pi}{6}$, the answer is *not* $\frac{5\pi}{6}$. First compute the tangent and then find a value in the domain of the inverse tangent whose tangent agrees. The reference angle for $\frac{5\pi}{6}$ is $\frac{\pi}{6}$ and $\frac{5\pi}{6}$ is in the third quadrant. In this quadrant the sine is positive and the cosine is negative, so

$$\tan\left(\frac{5\pi}{6}\right) = -\tan\left(\frac{\pi}{6}\right) = -\frac{\sin\left(\frac{\pi}{6}\right)}{\cos\left(\frac{\pi}{6}\right)} = -\frac{\frac{1}{2}}{\frac{\sqrt{3}}{2}} = -\frac{1}{\sqrt{3}} = -\frac{\sqrt{3}}{3}.$$

Since $\tan\left(-\frac{\pi}{6}\right)$ is also $-\frac{\sqrt{3}}{3}$,

$$\arctan\left(\tan\left(\frac{5\pi}{6}\right)\right) = -\frac{\pi}{6}.$$

■

**Example 162** *Find the exact value of the expression.*
(a) $\sin\left(\arccos\left(\frac{4}{5}\right)\right)$
(b) *(Exercise Set 3.7, Exercise 27)* $\cos\left(\arcsin\left(\frac{3}{5}\right) + \arccos\left(\frac{4}{5}\right)\right)$

Solution:
(a) Let $t$ satisfy

$$t = \arccos\left(\frac{4}{5}\right) \Rightarrow \cos t = \frac{4}{5}.$$

By the Pythagorean Identity,

$$(\cos t)^2 + (\sin t)^2 = 1$$
$$\left(\frac{4}{5}\right)^2 + (\sin t)^2 = 1$$
$$(\sin t)^2 = 1 - \frac{16}{25} = \frac{9}{25}$$
$$\sin t = \pm\sqrt{\frac{9}{25}} = \pm\frac{3}{5}.$$

Since $t = \arccos\left(\frac{4}{5}\right)$ and $t$ is in the interval $[0, \pi]$, where the sine is always nonnegative, we have

$$\sin t = \sin\left(\arccos\left(\frac{4}{5}\right)\right) = \frac{3}{5}.$$

(b) To simplify the expression first apply the sum formula for cosine, $\cos(a+b) = \cos a \cos b - \sin a \sin b$. This gives

$$\cos\left(\arcsin\left(\frac{3}{5}\right) + \arccos\left(\frac{4}{5}\right)\right)$$
$$= \cos\left(\arcsin\left(\frac{3}{5}\right)\right) \cos\left(\arccos\left(\frac{4}{5}\right)\right)$$
$$- \sin\left(\arcsin\left(\frac{3}{5}\right)\right) \sin\left(\arccos\left(\frac{4}{5}\right)\right)$$
$$= \cos\left(\arcsin\left(\frac{3}{5}\right)\right)\left(\frac{4}{5}\right) - \left(\frac{3}{5}\right)\sin\left(\arccos\left(\frac{4}{5}\right)\right).$$

The figure shows a triangle with angle $\theta$ satisfying $\sin\theta = \frac{3}{5}$.

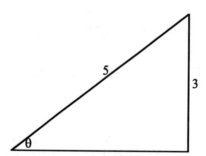

By the Pythagorean Theorem, the missing side has length

$$\sqrt{5^2 - 3^2} = \sqrt{25 - 9} = \sqrt{16} = 4,$$

so $\cos\theta = \frac{4}{5}$. Then

$$\frac{4}{5} = \cos\theta = \cos\left(\arcsin\left(\frac{3}{5}\right)\right)$$

and

## 3.7. INVERSE TRIGONOMETRIC FUNCTIONS

$$\frac{3}{5} = \sin\theta = \sin\left(\arccos\left(\frac{4}{5}\right)\right).$$

Finally,

$$\begin{aligned}\cos\left(\arcsin\left(\frac{3}{5}\right) + \arccos\left(\frac{4}{5}\right)\right) &= \frac{4}{5}\cdot\frac{4}{5} - \frac{3}{5}\cdot\frac{3}{5} \\ &= \frac{16}{25} - \frac{9}{25} \\ &= \frac{7}{25}.\end{aligned}$$

∎

**Example 163** *Solve the equation on the given interval, express the solution for x in terms of inverse functions, and use a calculator to approximate the solutions.*
   *(a) $\sin 2x - \cos x = 0$ on $\left[0, \frac{\pi}{2}\right]$*
   *(b) (Exercise Set 3.7, Exercise 29) $(\tan x)^2 - \tan x - 2 = 0$ on $\left(-\frac{\pi}{2}, \frac{\pi}{2}\right)$*

Solution:
(a) Use the double-angle formula for sine, $\sin 2x = 2\sin x \cos x$.

$$\begin{aligned}\sin 2x - \cos x &= 0 \\ 2\sin x \cos x - \cos x &= 0 \\ \cos x(2\sin x - 1) &= 0 \\ \cos x &= 0, \ \sin x = \frac{1}{2}\end{aligned}$$

so

$$x = \arccos(0) = \frac{\pi}{2} \approx 1.571, \ x = \arcsin\left(\frac{1}{2}\right) = \frac{\pi}{6} \approx 0.524$$

(b) If we let $u = \tan x$, then

$$\begin{aligned}(\tan x)^2 - \tan x - 2 &= u^2 - u - 2 \\ &= (u-2)(u+1).\end{aligned}$$

So

$$(\tan x)^2 - \tan x - 2 = 0$$
$$(\tan x - 2)(\tan x + 1) = 0$$
$$\tan x = 2 \text{ or } \tan x = -1$$

and

$$x = \arctan(2) \approx 1.107, \text{ or}$$
$$x = \arctan(-1) = -\frac{\pi}{4} \approx -0.785.$$

∎

**Example 164** *(Exercise Set 3.7, Exercise 33)* Verify the identity

$$\arccos x = \arcsin(\sqrt{1-x^2}),$$

where $|x| \leq 1$.

Solution: The definition of arccos is,

$$y = \arccos x \Leftrightarrow \cos y = x.$$

The figure shows a right triangle with one angle $y$ and $\cos y = x$. By the Pythagorean Theorem, the missing side $z$ in the figure is

$$1 = x^2 + z^2$$
$$z = \sqrt{1-x^2}.$$

Then

$$\sin y = \sqrt{1-x^2} \Leftrightarrow y = \arcsin\left(\sqrt{1-x^2}\right)$$

and

$$\arccos x = \arcsin\left(\sqrt{1-x^2}\right).$$

## 3.7. INVERSE TRIGONOMETRIC FUNCTIONS

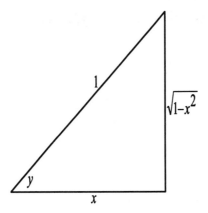

■

**Example 165** *(Exercise Set 3.7, Exercise 38) A lighthouse is 4 miles from a straight shoreline as shown in the figure. If the light from the lighthouse is moving along the shoreline, express the angle $\theta$ formed by the beam of light and the shoreline in terms of the distance $x$ in the figure.*

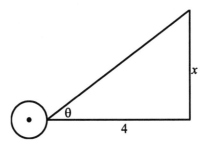

Solution: The parameters $\theta, x$, and 4 can be related by a tangent function since the sides of length $x$ and 4 are the opposite and adjacent sides of the angle $\theta$. So
$$\tan \theta = \frac{x}{4}$$
and
$$\theta = \arctan\left(\frac{x}{4}\right).$$

■

## 3.8 Applications of Trigonometric Functions

### 3.8.1 Law of Cosines

To *solve* a triangle, that is, find the lengths of all sides and all angles, the *Law of Cosines* can be used when we know two sides and the angle between them or all three sides.

<u>Law of Cosines</u>

$$a^2 = b^2 + c^2 - 2bc \cos \alpha$$
$$b^2 = a^2 + c^2 - 2ac \cos \beta$$
$$c^2 = a^2 + b^2 - 2ab \cos \gamma$$

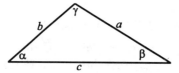

**Example 166** *(Exercise Set 3.8, Exercise 1)* Let the angles of a triangle be $\alpha, \beta$, and $\gamma$ with opposite sides $a, b,$ and $c$, respectively. If $\alpha = 35°, b = 15,$ and $c = 25$, use the Law of Cosines to solve the triangle, rounding all answers to one decimal place.

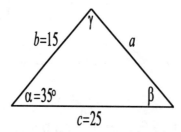

Solution: We need to find the missing parts $\beta, \gamma,$ and $a$. Since we are given the two sides $b$ and $c$ and the angle $\alpha$ between them, the side $a$ can be found. That is,

$$a^2 = b^2 + c^2 - 2bc \cos \alpha$$
$$a^2 = 15^2 + 25^2 - 2(15)(25) \cos 35°$$
$$= 225 + 625 - 750 \cos 35°$$
$$a = \sqrt{850 - 750 \cos 35°} \approx 15.4.$$

## 3.8. APPLICATIONS OF TRIGONOMETRIC FUNCTIONS

Now that we have an approximation to side $a$, we can find either of the angles $\beta$ or $\gamma$, since we have all three sides. We elect to find $\beta$, so

$$b^2 = a^2 + c^2 - 2ac\cos\beta$$
$$15^2 = (15.4)^2 + 25^2 - 2(15.4)(25)\cos\beta$$
$$225 = 237.16 + 625 - 770\cos\beta$$
$$\cos\beta = \frac{225 - 862.16}{-770}$$
$$= \frac{637.16}{770}$$
$$\beta = \arccos\left(\frac{637.16}{770}\right) \approx 0.5962.$$

In degrees we have

$$\beta \approx 0.5962\left(\frac{180}{\pi}\right) \approx 34.2°$$

Since the sum of the angles of a triangle is 180°,

$$\gamma \approx 180 - 35 - 34.2 = 110.8°.$$

■

**Example 167** *(Exercise Set 3.8, Exercise 19)* A gas pipeline is to be constructed between towns A and B. Engineers have two alternatives. They can connect A and B directly, but then they must build the pipeline through a swamp, or they can build the pipeline from town A to town C, which is 3 miles directly west of A, and then to town B which is 2 miles directly northwest of C. The cost of construction through the swamp from A to B is $125,000.00 per mile and the cost to go through C is $100,000.00 per mile.

(a) Which alternative should the engineers select?

(b) At what cost per mile for construction through C would there be no price difference in the two alternatives?

Solution: An illustration of the situation is shown below.

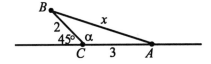

(a) To determine the cost per mile we need to find the distance from $A$ to $B$. If we know the angle $\alpha$ shown in the figure, then we could use the Law of Cosines, since we would have the angle opposite the length we need and the two sides adjacent the angle. Since point $B$ is directly northwest of point $C$, the angle made by the line segment connecting the two points is $45°$. So

$$\alpha = 180 - 45 = 135°.$$

By the Law of Cosines

$$\begin{aligned}x^2 &= 2^2 + 3^2 - 2(2)(3)\cos 135° \\ x^2 &= 13 - 12\cos 135° \\ x &= \sqrt{13 - 12\cos 135°} \approx 4.6 \text{ miles.}\end{aligned}$$

The cost of construction directly between points $A$ and $B$ is

$$\left(\sqrt{13 - 12\cos 135°}\right)(125,000) \approx \$579,403.00.$$

The cost of construction from $A$ to $C$ and then from $C$ to $B$ is

$$(3+2)(100,000) = \$500,000.00.$$

So it is cheaper to avoid the swamp.

(b) Let $P$ be the cost per mile for construction through $C$. If the total cost of construction from $A$ to $C$ to $B$ is to equal the cost directly from $A$ to $B$, through the swamp, then

$$\begin{aligned}5P &\approx 579,403 \\ P &\approx \$115,881.00.\end{aligned}$$

∎

### 3.8.2 Law of Sines

The *Law of Sines* states that the lengths of the sides of a triangle are proportional to the corresponding opposite angles. In a triangle with sides $a, b$ and $c$ having corresponding opposite angles $\alpha, \beta$ and $\gamma$, the Law of Sines states that

$$\frac{\sin \alpha}{a} = \frac{\sin \beta}{b} = \frac{\sin \gamma}{c}.$$

The Law of Sines can be used to solve a triangle when you have a side and the angle opposite it.

## 3.8. APPLICATIONS OF TRIGONOMETRIC FUNCTIONS

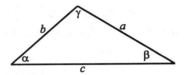

**Example 168** *(Exercise Set 3.8, Exercise 8)* *Let the angles of a triangle be $\alpha, \beta,$ and $\gamma$ with opposite sides $a, b,$ and $c$, respectively. If $\alpha = 60°, \beta = 56°$, and $a = 25$ use the Law of Sines to solve the triangle, rounding all answers to one decimal place.*

Solution: We need to find the missing parts $\gamma, b$ and $c$. Since the sum of the angles of a triangle is $180°$ and we are given two angles, $\alpha$ and $\beta$, we can find $\gamma$ immediately by

$$\gamma = 180 - 60 - 56 = 64°.$$

To find $b$ we have

$$\frac{\sin \beta}{b} = \frac{\sin \alpha}{a}$$
$$b = \frac{a \sin \beta}{\sin \alpha} = \frac{25 \sin 56°}{\sin 60°} \approx 23.9.$$

To find $c$ we have

$$\frac{\sin \gamma}{c} = \frac{\sin \alpha}{a}$$
$$c = \frac{a \sin \gamma}{\sin \alpha} = \frac{25 \sin 64°}{\sin 60°} \approx 25.9.$$

■

**Example 169** *(Exercise Set 3.8, Exercise 15)* *Show there is no triangle satisfying the conditions $a = 3, b = 10,$ and $\alpha = 25.4°$.*

Solution: If there was such a triangle, then the angle $\beta$ could be found by the Law of Sines from

$$\frac{\sin \beta}{b} = \frac{\sin \alpha}{a}$$
$$\sin \beta = \frac{b \sin \alpha}{a} = \frac{10 \sin 25.4°}{3} \approx 1.43.$$

Since the sine of an angle is never greater than one, we must conclude that no such triangle exists.

■

**Example 170** *(Exercise Set 3.8, Exercise 23) The National Forest Service maintains observation towers to check for the outbreak of forest fires. Suppose two towers are at the same elevation, one at point A and another 10 miles due west at a point B. The ranger at A spots a fire in the northwest whose line of sight makes an angle of 63° with the line between the towers, and contacts the ranger at B. This ranger locates the fire along a line of sight that makes a 50° angle with the line of the towers. How far is the fire from tower B?*

Solution:

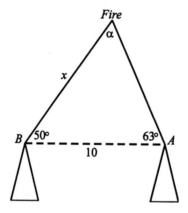

If we knew the angle $\alpha$, then the Law of Sines would yield

$$\frac{\sin\alpha}{10} = \frac{\sin 63°}{x}$$

$$x = \frac{10\sin 63°}{\sin\alpha}.$$

Since

$$\alpha = 180 - 50 - 63 = 67°$$

we have

$$x = \frac{10\sin 63°}{\sin 67°} \approx 9.7.$$

So the fire is approximately 9.7 miles from tower $B$.
∎

## 3.8.3 Heron's Formula

Given the lengths of the three sides of a triangle an easy formula for finding the area of the triangle is *Heron's formula*. A triangle with sides of lengths $a, b$ and $c$ has area

$$A = \sqrt{s(s-a)(s-b)(s-c)}, \text{ where } s = \frac{1}{2}(a+b+c).$$

**Example 171** *Find the area of a triangle with sides of lengths 12 ft, 18 ft and 24 ft.*

Solution: To use Heron's formula first compute half the perimeter of the triangle. So,

$$\begin{aligned} s &= \frac{1}{2}(a+b+c) \\ &= \frac{1}{2}(12+18+24) \\ &= \frac{1}{2}(54) = 27. \end{aligned}$$

Then the area is

$$\begin{aligned} A &= \sqrt{27(27-12)(27-18)(27-24)} \\ &= \sqrt{27(15)(9)(3)} \\ &= \sqrt{10935} \\ &= 27\sqrt{15} \\ &\approx 104.57 \text{ ft}^2. \end{aligned}$$

■

**Example 172** *(Exercise Set 3.8, Exercise 24) The lengths of the sides of a triangular parcel of land are approximately 200 ft, 300 ft, and 450 ft. If the land is valued at $2,000.00 per acre, what is the value of the parcel of land?*

Solution: To determine how many acres in the parcel, we need to know the number of square feet of land in the parcel and also the number of square feet in one acre. The number of square feet in one acre is

$$\text{one acre} = 43,560 \text{ ft}^2.$$

The area can be found using Heron's formula. Since

$$\begin{aligned} s &= \frac{1}{2}(200 + 300 + 450) \\ &= \frac{1}{2}(950) = 475, \end{aligned}$$

we have

$$\begin{aligned} A &= \sqrt{475(475 - 200)(475 - 300)(475 - 450)} \\ &= \sqrt{475(275)(175)(25)} = \sqrt{571484375} = 625\sqrt{1463} \\ &\approx 23,906 \text{ ft}^2. \end{aligned}$$

The number of acres in the parcel is approximately

$$\frac{23906}{43560} = 0.55 \text{ acres}$$

and the value of the parcel of land is approximately

$$\left(\frac{23906}{43560}\right) 2000 \approx \$1,098.00.$$

■

# Chapter 4

# Exponential and Logarithm Functions

## 4.1 Introduction

The exponential functions extend the notion of exponent to include all real numbers. The *natural exponential* function $f(x) = e^x$ and the *natural logarithm* function $g(x) = \ln x$ are perhaps the most important function-inverse pair in science and mathematics.

## 4.2 The Natural Exponential Function

For $a > 0$, the *exponential function with base* $a$ is defined as $f(x) = a^x$ for all real numbers $x$. The value of the base $a$ determines the general shape of the graph of the exponential function.

$f(x) = a^x, a > 1$:
domain: $(-\infty, \infty)$
range: $(0, \infty)$
graph increasing: for all $x$
graph decreasing: for no $x$
horizontal asymptote: $y = 0$

$$x \to -\infty \Rightarrow f(x) \to 0$$
$$x \to \infty \Rightarrow f(x) \to \infty$$

y-intercept : $(0,1)$
x-intercept : none

$\underline{f(x) = a^x, 0 < a < 1}$ :
domain: $(-\infty, \infty)$
range: $(0, \infty)$
graph increasing: for no $x$
graph decreasing: for all $x$
horizontal asymptote: $y = 0$

$$x \to -\infty \Rightarrow f(x) \to \infty$$
$$x \to \infty \Rightarrow f(x) \to 0$$

y-intercept : $(0,1)$
x-intercept : none

$\underline{f(x) = a^x = 1, \text{ for } a = 1}$ : constant function

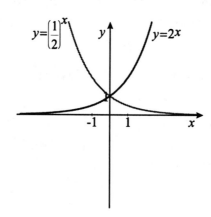

## 4.2.1 Graphs of Exponential Functions

**Example 173** *Use the graph of $y = 2^x$ shown in the figure to sketch the graph of the function.*
   (a) $y = 2^{(x-1)} + 1$   (b) $y = -2^{(x+1)} - 2$

## 4.2. THE NATURAL EXPONENTIAL FUNCTION

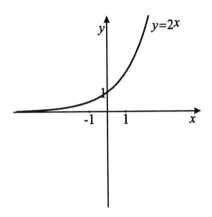

Solution: The shifting and scaling properties allow us to sketch reasonable graphs quickly once we know the general shape of the basic curve.

(a) The basic function is $y = 2^x$, so if the argument is changed to $(x-1)$ the graph of $y = 2^{(x-1)}$, is obtained by shifting the basic curve 1 unit to the right. To obtain the graph of $y = 2^{(x-1)} + 1$ from the graph of $y = 2^x$, first shift the basic curve 1 unit to the right and then shift the resulting curve 1 unit upward.

<u>y-intercept</u> : Set $x = 0$.

$$\begin{aligned} y &= 2^{-1} + 1 \\ &= \frac{1}{2} + 1 = \frac{3}{2} \end{aligned}$$

<u>Horizontal asymptote:</u> $y = 1$ since the horizontal asymptote of $y = 2^x$ is the line $y = 0$ and the graph is shifted 1 unit upward.

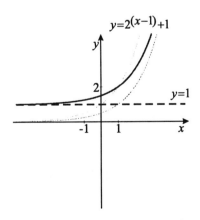

(b) The graph of $y = -2^x$ is the reflection of the graph of $y = 2^x$ about the $x$-axis, since introducing the minus sign reverses the signs of the $y$-coordinates of points on the curve. Changing the argument from $x$ to $x+1$ shifts the reflected graph 1 unit to the left. Finally, shift this graph 2 units downward to obtain the graph of $y = -2^{(x+1)} - 2$.

$\underline{y\text{-intercept}}$ : Set $x = 0$.

$$\begin{aligned} y &= -2^1 - 2 \\ &= -4 \end{aligned}$$

$\underline{\text{Horizontal asymptote:}}$ $y = -2$, since $y = -2^{(x+1)}$ was shifted 2 units downward.

Also $y = 2^x$ approaches the horizontal line $y = 0$ from above. The graph of $y = -2^{(x+1)} - 2$ was obtained from $y = 2^x$ by first reflecting it about the $x$-axis, so it will approach the line $y = -2$ from below.

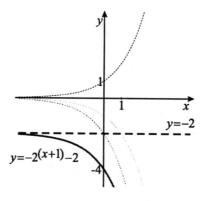

■

**Example 174** *Use the graph of $y = \left(\frac{1}{2}\right)^x$ shown in the figure to sketch the graph of $y = 3 \cdot 2^{(1-x)} + 1$.*

## 4.2. THE NATURAL EXPONENTIAL FUNCTION

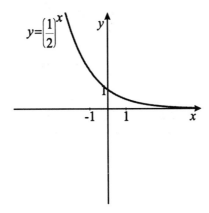

Solution: First note that

$$2^{-x} = \frac{1}{2^x} = \left(\frac{1}{2}\right)^x$$

so

$$\begin{aligned} y &= 3 \cdot 2^{(1-x)} + 1 \\ &= 3 \cdot 2^{-(x-1)} + 1 \\ &= 3 \cdot \left(\frac{1}{2}\right)^{x-1} + 1. \end{aligned}$$

Start with the graph of $y = \left(\frac{1}{2}\right)^x$, vertically stretch the graph by a factor of 3, shift the resulting graph 1 unit to the right, and 1 unit upward to obtain the graph of $y = 3 \cdot 2^{(1-x)} + 1$.

<u>y-intercept</u> : Set $x = 0$.

$$\begin{aligned} y &= 3 \cdot 2^1 + 1 \\ &= 7 \end{aligned}$$

<u>Horizontal asymptote:</u> $y = 1$

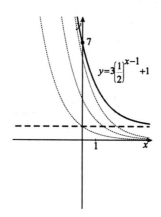

■

## 4.2.2 The Natural Exponential Function

As $n \to \infty$ the real numbers $\left(1 + \frac{1}{n}\right)^n$ approach the irrational number $e \approx 2.71828$. The function
$$f(x) = e^x$$
is called the *natural exponential function*.

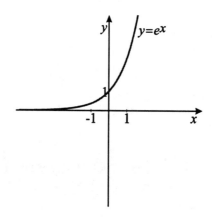

**Example 175** *Sketch the graph of the function.*
  (a) *(Exercise Set 4.2, Exercise 8)*   $f(x) = 3 - e^{-(x-1)}$
  (b) *(Exercise Set 4.2, Exercise 12)*   $f(x) = e^{-|x|}$

Solution:
  (a) For any function $f$, if the argument $x$ is replaced with $-x$, the graph of $y = f(-x)$ is the reflection about the y-axis of $y = f(x)$.

## 4.2. THE NATURAL EXPONENTIAL FUNCTION

Start with the graph of $y = e^x$.
1. Graph of $y = e^{-x}$ : Reflect the graph of $y = e^x$ about the $y$-axis.
2. Graph of $y = -e^{-x}$ : Reflect the graph of $y = e^{-x}$ about the $x$-axis.
3. Graph of $y = -e^{-(x-1)}$ : Shift the graph of $y = -e^{-x}$ to the right 1 unit.
4. Graph of $y = 3 - e^{-(x-1)}$ : Shift the graph of $y = -e^{-(x-1)}$ upward 3 units.

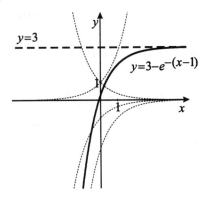

(b) Replacing $x$ with $|x|$ changes the graph of $y = e^{-x}$ to one that is symmetric with respect to the $y$-axis. Since

$$e^{-|x|} = e^{-|-x|}$$

For example, if $x > 0$, the points with first coordinate $x$ and first coordinate $-x$ have the same $y$-coordinates.

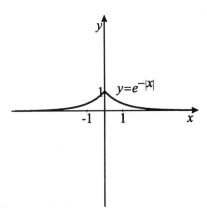

## 192  CHAPTER 4. EXPONENTIAL AND LOGARITHM FUNCTIONS

**Example 176** *(Exercise Set 4.2, Exercise 16) Use a graphing device to approximate all solutions to the equation $xe^x = x^2 + 4x + 2$.*

Solution: To approximate the solutions, plot $y = xe^x$ and $y = x^2 + 4x + 2$ together on the same coordinate axes and determine the points of intersection. When using a graphing device, it is essential to have a viewing rectangle that shows the important features. A viewing rectangle of $[-10, 10] \times [-5, 20]$ shows three points of intersection. A viewing rectangle of $[-5, 5] \times [-5, 5]$ would show only two of those points. From the initial plot, the points of intersection occur at approximately $x = -3.5$, $x = -0.8$ and $x = 1.9$. Zooming in near these points we get better accuracy with the points of intersection being approximately,

$$(-3.38, -0.12), (-0.72, -0.35), (1.94, 13.5).$$

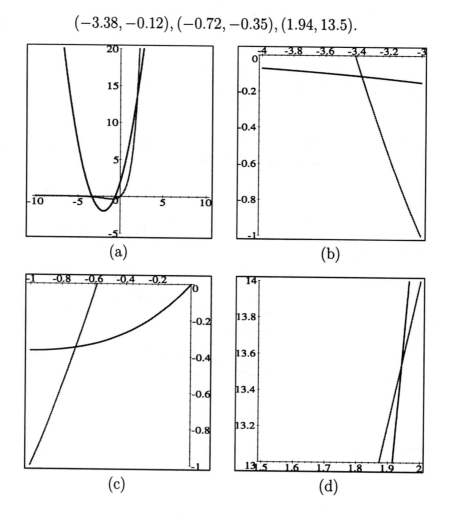

■

## 4.2. THE NATURAL EXPONENTIAL FUNCTION

**Example 177** *(Exercise Set 4.2, Exercise 24) Use a graphing device to compare the rates of growth of $f(x) = e^x$ and $g(x) = x^{10}$ by graphing the functions together in several appropriate viewing rectangles. Approximate the solutions to $e^x = x^{10}$.*

Solution: We want to determine whether, for large $x$,
$$x^{10} > e^x \quad \text{or} \quad e^x > x^{10}.$$

We will sketch the graph of $y = x^{10}$ and $y = e^x$. Whichever is eventually above the other, produces the dominant function.

Since both functions are nonnegative for all $x$, there is no need to show much of the negative $y$-axis. The first viewing rectangle we choose is
$$[-5, 5] \times [-1, 5].$$

In this view it appears that $g(x) = x^{10}$ eventually grows faster than $f(x) = e^x$. We should not jump to a quick conclusion. For a second viewing rectangle we select
$$[0, 20] \times [0, 10^5].$$

It still appears that $g(x) = x^{10}$ grows faster than $f(x) = e^x$. One last viewing rectangle reveals another story. Try
$$[0, 50] \times [0, 10^{16}].$$

This time we see that $f(x) = e^x$ eventually overtakes $g(x) = x^{10}$ and in fact eventually grows much faster.

The figures show points of intersection occurring when
$$x \approx -0.9, \ x \approx 1.1, \ \text{and} \ x \approx 35.6$$

which are the approximate solutions to the equation $e^x = x^{10}$.

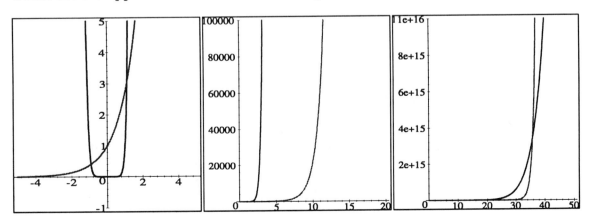

### 4.2.3 Solving Equations Involving $e^x$

**Example 178** *Solve the equation.*

(a) $2x^3 e^x + x^4 e^x = 0$    (b) $x^2 e^x - 2xe^x = 3e^x$

Solution:

(a) Since $e^x$ is common to both terms, it can be factored from the terms. So

$$e^x(2x^3 + x^4) = 0.$$

Since $e^x > 0$ for all $x$, both sides of the equation can be divided by $e^x$ and the last equation will be 0, only when

$$\begin{aligned} 2x^3 + x^4 &= 0 \\ x^3(2+x) &= 0 \\ x &= 0, \quad x = -2. \end{aligned}$$

(b) First rewrite the equation so that one side is 0.

$$\begin{aligned} x^2 - 2x - 3 &= 0 \\ (x-3)(x+1) &= 0 \\ x &= 3, \quad x = -1 \end{aligned}$$

Since $e^x > 0$ we have

$$\begin{aligned} x^2 e^x - 2xe^x &= 3e^x \\ x^2 e^x - 2xe^x - 3e^x &= 0 \\ e^x(x^2 - 2x - 3) &= 0. \end{aligned}$$

### 4.2.4 Compound Interest

If an initial amount of $A_0$ dollars is invested at an interest rate $i$ compounded $n$ times a year, the investment after $t$ years has a value

$$A_n(t) = A_0 \left(1 + \frac{i}{n}\right)^{nt}$$

## 4.2. THE NATURAL EXPONENTIAL FUNCTION

dollars. If the interest is compounded *continuously*, then the amount after $t$ years is

$$A_c(t) = A_0 e^{it}$$

dollars.

**Example 179** *(Exercise Set 4.2, Exercise 28) Suppose $1,000.00 is invested at 10% interest and the interest rate remains fixed for 8 years. Determine the value of the investment if the interest is compounded annually, semiannually, quarterly, monthly, weekly, daily, hourly and continuously.*

Solution:

| Interest Compounded | Times Interest Computed per Year | Value After 8 Years |
|---|---|---|
| Annually | 1 | $A_1(8) = \$2143.59$ |
| Semiannually | 2 | $A_2(8) = \$2182.88$ |
| Quaterly | 4 | $A_4(8) = \$2203.76$ |
| Monthly | 12 | $A_{12}(8) = \$2218.18$ |
| Weekly | 52 | $A_{52}(8) = \$2223.83$ |
| Daily | 365 | $A_{365}(8) = \$2225.30$ |
| Hourly | $24 \cdot 365 = 8760$ | $A_{8760}(8) = \$2225.53$ |
| Continuously | | $A_c(8) = \$2225.54$ |

The values in the table are computed as follows:
Annually: $A_1(8) = 1000(1 + 0.1)^8 = \$2143.59$
Semiannually: $A_2(8) = 1000 \left(1 + \frac{0.1}{2}\right)^{16} = \$2182.88$
Quarterly: $A_4(8) = 1000 \left(1 + \frac{0.1}{4}\right)^{32} = \$2203.76$
Monthly: $A_{12}(8) = 1000 \left(1 + \frac{0.1}{12}\right)^{96} = \$2218.18$
Weekly: $A_{52}(8) = 1000 \left(1 + \frac{0.1}{52}\right)^{416} = \$2223.83$
Daily: $A_{365}(8) = 1000 \left(1 + \frac{0.1}{365}\right)^{2920} = \$2225.30$
Hourly: There are 24 hours in a day and 365 days in the year so the interest is computed $24 \cdot 365 = 8760$ times.
$A_{8760}(8) = 1000 \left(1 + \frac{0.1}{8760}\right)^{70080} = \$2225.53$
Continuously: $A_c(8) = 1000 e^{0.8} = \$2225.54$
■

## 4.3 Logarithm Functions

For $a \neq 1$, the inverse function of the exponential function to the base $a$, $f(x) = a^x$, is called the *logarithm function to the base $a$*, and written $g(x) = \log_a x$. The functions are then related by the following relations:

For each $x$ in $(0, \infty)$ we have

$$y = \log_a x \Leftrightarrow x = a^y.$$

As a consequence, for each $x$ in $(0, \infty)$ and for each real number $y$ we have

$$\log_a a^y = y, \quad \text{and} \quad a^{\log_a x} = x.$$

Recall that the domain of the exponential function is the set of all real numbers so the range of the logarithm function is also the set of all real numbers. The range of the exponential function is $(0, \infty)$, so the domain of the logarithm function is $(0, \infty)$.

The inverse to the natural exponential function $f(x) = e^x$ is the *natural logarithm function*. The natural logarithm function is the logarithm to the base $e$, written $y = \log_e x = \ln x$. So we have the following:

For each $x$ in $(0, \infty)$

$$y = \ln x \Leftrightarrow x = e^y.$$

For each $x$ in $(0, \infty)$ and each real number $y$

$$\ln e^y = y \quad \text{and} \quad e^{\ln x} = x.$$

### 4.3.1 Evaluation of Logarithms

**Example 180** *Evaluate the expression.*
(a) $\log_2 64$  (b) $\log_4 16$  (c) $\log_{1/3} 3$  (d) $\log_3 \frac{1}{27}$
(e) $e^{\ln 6}$  (f) $\ln e^{\sqrt{2}}$

Solution:
(a) Using the inverse relation between the logarithm function and the exponential function gives

$$x = \log_2 64 \Leftrightarrow 2^x = 64$$
$$x = 6.$$

## 4.3. LOGARITHM FUNCTIONS

(b)
$$x = \log_4 16 \Leftrightarrow 4^x = 16$$
$$x = 2$$

(c)
$$x = \log_{1/3} 3 \Leftrightarrow \left(\frac{1}{3}\right)^x = 3$$
$$x = -1$$

(d)
$$x = \log_3 \frac{1}{27} \Leftrightarrow 3^x = \frac{1}{27}$$
$$x = -3$$

since $3^{-3} = \frac{1}{3^3} = \frac{1}{27}$.

(e) The exponential and logarithm functions are inverses of one another and therefore each undoes the process of the other. That is, if a value is input to ln and the resulting number then used as input to $e$, the final output is the original value.

$$x \to \boxed{\ln\Box} \to \boxed{e^{\ln x}} \to x$$

So
$$e^{\ln 6} = 6.$$

(f)
$$\ln e^{\sqrt{2}} = \sqrt{2}$$

∎

**Example 181** *Solve the equation.*
(a) $\log_3(2x - 5) = 2$
(b) *(Exercise Set 4.3, Exercise 38)* $\log_2(5x^2 - 8x) = 2$
(c) $e^{2x-1} = 2$

Solution:
(b)

$$\begin{aligned} \log_3(2x-5) &= 2 \\ 2x-5 &= 3^2 \\ 2x &= 14 \\ x &= 7 \end{aligned}$$

(b)

$$\begin{aligned} \log_2(5x^2-8x) &= 2 \\ 5x^2-8x &= 2^2 \\ 5x^2-8x &= 4 \end{aligned}$$

To solve the quadratic rewrite the expression with one side 0 and factor. So

$$\begin{aligned} 5x^2-8x &= 4 \\ 5x^2-8x-4 &= 0 \\ (5x+2)(x-2) &= 0 \\ 5x+2 &= 0, \text{ or } x-2=0 \\ x &= -\frac{2}{5}, \quad x=2. \end{aligned}$$

(c) The inverse relationship between the exponential and logarithm functions is the key here. If we take the natural logarithm of the left side the result is the input to the natural exponential function, $2x-1$. So as not to change the equation, we take the natural logarithm of both sides and simplify. So

$$\begin{aligned} e^{2x-1} &= 2 \\ \ln e^{2x-1} &= \ln 2 \\ 2x-1 &= \ln 2 \\ 2x &= 1+\ln 2 \\ x &= \frac{1+\ln 2}{2}. \end{aligned}$$

■

## 4.3.2 Arithmetic Properties of Logarithms

There are three important properties of the logarithm functions.

$$\log_a (x_1 x_2) = \log_a x_1 + \log_a x_2$$
$$\log_a \left(\frac{x_1}{x_2}\right) = \log_a x_1 - \log_a x_2$$
$$\log_a x_1^r = r \log_a x_1$$

For emphasis we also list the properties for the natural logarithm function.

$$\ln (x_1 x_2) = \ln x_1 + \ln x_2$$
$$\ln \left(\frac{x_1}{x_2}\right) = \ln x_1 - \ln x_2$$
$$\ln x_1^r = r \ln x_1$$

**Example 182** *Use the properties of logarithms to simplify the expression so that the result does not contain logarithms of products, quotients, or powers.*

(a) $\ln x(x^2 + 1)$
(b) (Exercise Set 4.3, Exercise 14)   $\ln \frac{1}{x}$
(c) (Exercise Set 4.3, Exercise 18)   $\ln \frac{x \sqrt[3]{x^2}}{(x+2)^3}$
(d) $\log_5 \sqrt{\frac{x^3}{4x^2-2}}$

Solution:
(a) The expression in the natural logarithm function is the product of the two terms $x$ and $(x^2 + 1)$, so the product rule can be used to give

$$\ln x(x^2 + 1) = \ln x + \ln(x^2 + 1).$$

This is the final answer! The expression $\ln(x^2 + 1)$ cannot be further simplified.

(b) Because $e^0 = 1$, we have $\ln 1 = 0$, and

$$\ln \frac{1}{x} = \ln 1 - \ln x$$
$$= -\ln x.$$

This can also be recognized using the power rule. That is,

$$\ln \frac{1}{x} = \ln x^{-1} = -\ln x.$$

(c) This problem may look complicated but just be careful to apply the properties in a proper sequence. The first property to apply is the quotient rule, since the entire expression is one quotient. Once the quotient rule is applied the two resulting pieces will be treated separately and the properties applied to them individually. To simplify the steps some, first rewrite

$$\sqrt[3]{x^2} = \left(x^2\right)^{1/3} = x^{\frac{2}{3}}.$$

Then we have

$$\ln \frac{xx^{2/3}}{(x+2)^3} = \ln x^{5/3} - \ln(x+2)^3$$
$$= \frac{5}{3}\ln x - 3\ln(x+2).$$

(d) First we write

$$\log_5 \sqrt{\frac{x^3}{4x^2 - 2}} = \log_5 \left(\frac{x^3}{4x^2 - 2}\right)^{1/2}$$
$$= \frac{1}{2}\log_5 \frac{x^3}{4x^2 - 2}$$
$$= \frac{1}{2}\left[\log_5 x^3 - \log_5(4x^2 - 2)\right]$$
$$= \frac{1}{2}\log_5 x^3 - \frac{1}{2}\log_5(4x^2 - 2)$$
$$= \frac{3}{2}\log_5 x - \frac{1}{2}\log_5(4x^2 - 2)$$

It may appear that the simplification is done, but sone further simplification is possible by recognizing that

$$(4x^2 - 2) = 4\left(x^2 - \frac{1}{2}\right)$$
$$= 4\left(x - \sqrt{\frac{1}{2}}\right)\left(x + \sqrt{\frac{1}{2}}\right)$$
$$= 4\left(x - \frac{\sqrt{2}}{2}\right)\left(x + \frac{\sqrt{2}}{2}\right).$$

## 4.3. LOGARITHM FUNCTIONS

So

$$\log_5 \sqrt{\frac{x^3}{4x^2-2}} = \frac{3}{2}\log_5 x - \frac{1}{2}\log_5(4x^2-2)$$

$$= \frac{3}{2}\log_5 x - \frac{1}{2}\log_5 4\left(x - \frac{\sqrt{2}}{2}\right)\left(x + \frac{\sqrt{2}}{2}\right)$$

$$= \frac{3}{2}\log_5 x - \frac{1}{2}\log_5 4 - \frac{1}{2}\log_5\left(x - \frac{\sqrt{2}}{2}\right)\left(x + \frac{\sqrt{2}}{2}\right)$$

$$= \frac{3}{2}\log_5 x - \frac{1}{2}\log_5 4 - \frac{1}{2}\log_5\left(x - \frac{\sqrt{2}}{2}\right) - \frac{1}{2}\log_5\left(x + \frac{\sqrt{2}}{2}\right).$$

Notice we used the product rule twice on the term

$$\frac{1}{2}\log_5 4\left(x - \frac{\sqrt{2}}{2}\right)\left(x + \frac{\sqrt{2}}{2}\right).$$

First on $\left[\left(x - \frac{\sqrt{2}}{2}\right)\left(x + \frac{\sqrt{2}}{2}\right)\right]$ and then on $\left[\left(x - \frac{\sqrt{2}}{2}\right)\left(x + \frac{\sqrt{2}}{2}\right)\right]$.
∎

**Example 183** *Solve the equation for $x$.*
 (a) (Exercise Set 4.3, Exercise 33)   $\ln x + \ln(x-1) = \ln 2$
 (b) (Exercise Set 4.3, Exercise 36)   $2\ln(x+2) - \ln x = \ln 8$
 (c) (Exercise Set 4.3, Exercise 39)   $e^{2x} = 3^{x-4}$

Solution:
(a) This time we use the properties of logarithms to combine the ln expressions into one expression. So

$$\begin{aligned}
\ln x + \ln(x-1) &= \ln 2 \\
\ln x(x-1) &= \ln 2 \\
x(x-1) &= 2 \\
x^2 - x - 2 &= 0 \\
(x-2)(x+1) &= 0 \\
x = 2, \quad x &= -1.
\end{aligned}$$

Since $\ln(x-1)$ is not defined at $x = 1$, the only true solution is $x = 2$ and $x = -1$ is an extraneous solution.

Note that $\ln x(x-1) = \ln 2$ gives $x(x-1) = 2$ for several reasons. We could use the fact that the function ln is one-to-one. Alternatively we could note that if each side is raised to an exponent of $e$, then $e^{\ln x(x-1)} = e^{\ln 2}$. By the inverse relation, this gives $x(x-1) = 2$.

(b) Using the logarithm rules

$$\begin{aligned}
2\ln(x+2) - \ln x &= \ln 8 \\
\ln(x+2)^2 - \ln x &= \ln 8 \\
\ln \frac{(x+2)^2}{x} &= \ln 8 \\
\frac{(x+2)^2}{x} &= 8 \\
x^2 + 4x + 4 &= 8x \\
x^2 - 4x + 4 &= 0 \\
(x-2)(x-2) &= 0 \\
x &= 2
\end{aligned}$$

(c) Taking the natural logarithm of both sides gives

$$\begin{aligned}
e^{2x} &= 3^{x-4} \\
\ln e^{2x} &= \ln 3^{x-4} \\
2x &= (x-4)\ln 3 \\
2x &= x\ln 3 - 4\ln 3 \\
2x - x\ln 3 &= -4\ln 3 \\
x(2 - \ln 3) &= \ln 3^{-4} \\
x &= \frac{\ln \frac{1}{81}}{2 - \ln 3} = \frac{\ln 81}{\ln 3 - 2}.
\end{aligned}$$

■

## 4.3.3 Graphs of Logarithm Functions

There are several important properties to remember when sketching graphs of logarithms. The most important graphs are for $a > 1$, which have the following properties:

<u>Domain:</u> $(0, \infty)$, so the graph of $y = \log_a x$ is on the right side of the $y$-axis

## 4.3. LOGARITHM FUNCTIONS

Range: $(-\infty, \infty)$ and

$$x \to \infty \Rightarrow \log_a x \to \infty.$$

Although the logarithm grows arbitrarily large as $x$ grows large, the growth is very slow. On the other hand exponential growth is very rapid.

Vertical Asymptote: $x = 0$ and

$$x \to 0^+ \Rightarrow \log_a x \to -\infty.$$

Notice that the logarithm function is not defined at $x = 0$, and since the domain is only positive real numbers, $x$ can only approach 0 from the right side. This is the reason for the + exponent on the 0.

x-intercept: $(1, 0)$, that is, the graph crosses the x-axis when $x = 1$, since

$$\log_a x = 0 \Leftrightarrow a^0 = x, \quad \text{so} \quad x = 1.$$

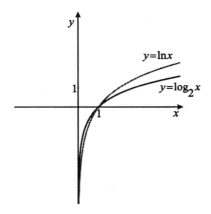

**Example 184** *Sketch the graph of the function.*
(a) $y = 2 - \ln(x - 1)$
(b) *(Exercise Set 4.3, Exercise 47)* $y = \ln(-x)$

Solution:
(a) To sketch the graph we will use the following steps.
1. Use $y = \ln x$ as the basic graph.
2. Use (1) to plot $y = -\ln x$.
3. Use (2) to plot $y = -\ln(x - 1)$.
4. Use (3) to plot $y = 2 - \ln(x - 1)$.

The graph of $y = -\ln x$ is obtained by reflecting the basic graph about the $x$-axis. So

$$x \to 0^+ \Rightarrow -\ln x \to \infty$$
$$x \to \infty \Rightarrow -\ln x \to -\infty.$$

Now shift $y = -\ln x$ to the right 1 unit to obtain the graph of $y = -\ln(x-1)$. Finally, shift the graph of $y = -\ln(x-1)$ upward 2 units to obtain the graph of $y = 2 - \ln(x-1)$.

Vertical asymptote: $x = 1$

$x$-intercept: Solve

$$\begin{aligned} 2 - \ln(x-1) &= 0 \Rightarrow \\ -\ln(x-1) &= -2 \\ \ln(x-1) &= 2 \\ e^{\ln(x-1)} &= e^2 \\ x - 1 &= e^2 \\ x &= e^2 + 1. \end{aligned}$$

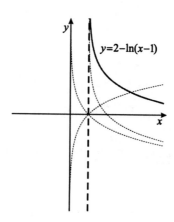

(b) The graph of $y = \ln(-x)$ is just the reflection of the graph of $y = \ln x$ about the $y$-axis.

## 4.4. EXPONENTIAL GROWTH AND DECAY

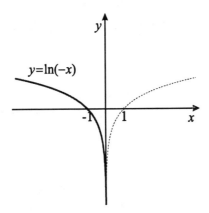

∎

**Example 185** *(Exercise Set 4.3, Exercise 58) Determine the length of time it takes an initial investment to triple in value if it earns 10% compounded continuously.*

Solution: The value of the investment after $t$ years is given by the exponential formula
$$A_c(t) = A_0 e^{0.1t},$$
where $A_0$ is some initial investment and 0.1 represents the 10% interest rate. To determine the length of time it takes the initial investment to triple solve for time $t$ in the equation

$$\begin{aligned} 3A_0 &= A_0 e^{0.1t} \\ 3 &= e^{0.1t} \\ \ln 3 &= \ln e^{0.1t} \\ \ln 3 &= 0.1t \\ t &= \frac{\ln 3}{0.1} = \frac{\ln 3}{\frac{1}{10}} = 10 \ln 3 \text{ years} \approx 11 \text{ years}. \end{aligned}$$

∎

## 4.4 Exponential Growth and Decay

If a quantity grows or decays at a rate that is directly proportional to the amount of the quantity that is present, then the quantity present at any time

$t$ can be modeled by an exponential function. If the initial amount of the quantity is $Q_0$, then the amount at any time $t$ is

$$Q(t) = Q_0 e^{kt}$$

where $k$ is the *constant of proportionality* that depends on the specific situation. If $k > 0$, we say $Q$ *grows exponentially* and if $k < 0$, we say $Q$ *decays exponentially*.

**Example 186** *(Exercise Set 4.4, Exercise 2) A bacteria culture starts with 500 bacteria and 5 hours later has 4,000 bacteria.*
(a) *Find an expression for the number of bacteria after $t$ hours.*
(b) *Find the number of bacteria that will be present after 6 hours.*
(c) *When will the population reach 15,000?*
(d) *How long does it take the population to double in size?*

Solution:
(a) Use the information given to find the specific values for $Q_0$ and $k$. Since the initial amount of bacteria is 500,

$$Q_0 = 500.$$

To find $k$, use the fact that after 5 hours, that is, when $t = 5$, there are 4,000 bacteria present. Then $Q(5) = 4000$, so

$$\begin{aligned} 4000 &= Q(5) = 500 e^{k(5)} \\ e^{5k} &= \frac{4000}{500} = 8 \\ \ln e^{5k} &= \ln 8 \\ 5k &= \ln 8 \\ k &= \frac{\ln 8}{5}. \end{aligned}$$

The number of bacteria after $t$ hours is

$$Q(t) = 500 e^{\frac{\ln 8}{5} t}.$$

(b) After $t = 6$ hours the number of bacteria present is

$$Q(6) = 500 e^{\frac{\ln 8}{5}(6)} \approx 6063.$$

## 4.4. EXPONENTIAL GROWTH AND DECAY

Notice that the expression for $Q(6)$ can be rewritten in the form

$$\begin{aligned} 500e^{\frac{\ln 8}{5}(6)} &= 500e^{(\ln 8)\frac{6}{5}} \\ &= 500\left(e^{\ln 8}\right)^{\frac{6}{5}} \\ &= 500(8)^{\frac{6}{5}}. \end{aligned}$$

(c) Find the value for time $t$ so that $Q(t) = 15,000$. Solve

$$Q(t) = 500e^{\frac{\ln 8}{5}t} = 15000,$$

then

$$e^{\frac{\ln 8}{5}t} = \frac{15000}{500} = 30$$

$$\ln e^{\frac{\ln 8}{5}t} = \ln 30$$

$$\frac{\ln 8}{5}t = \ln 30$$

and

$$t = \frac{5\ln 30}{\ln 8} \approx 8.2 \text{ hours.}$$

(d) Find the value of time $t$ so that $Q(t) = 2(500) = 1000$. Solve

$$Q(t) = 500e^{\frac{\ln 8}{5}t} = 1000,$$

then

$$e^{\frac{\ln 8}{5}t} = 2$$

$$\frac{\ln 8}{5}t = \ln 2$$

and

$$t = \frac{5\ln 2}{\ln 8} \approx 1.7 \text{ hours.}$$

∎

**Example 187** *(Exercise Set 4.4, Exercise 4)* The radioactive isotope thorium-234 has a half-life of approximately 578 hours.

(a) If a sample has a mass of 50 mg, find an expression for the mass after $t$ hours.

(b) How much will remain after 100 hours?

(c) When will the mass decay to 10 mg?

(d) Use a graphing device to sketch the graph of the mass function.

Solution:

(a) The *half-life* of a radioactive substance is the amount of time it takes for one half of the substance to decay. To find the constant of proportionality, use the fact that the half-life is 578 hours, so that at time $t = 578$ hours the mass of the substance will be 25 mg. So

$$Q(t) = 50e^{kt}$$

that is,

$$25 = Q(578) = 50e^{578k}.$$

So

$$e^{578k} = \frac{25}{50} = \frac{1}{2}$$
$$578k = \ln\frac{1}{2} = \ln 1 - \ln 2$$
$$= -\ln 2$$
$$k = -\frac{\ln 2}{578}$$

and

$$Q(t) = 50e^{-\frac{\ln 2}{578}t}.$$

Note that we used the arithmetic property of logarithms,

$$\ln\left(\frac{x}{y}\right) = \ln x - \ln y$$

and the fact that $\ln 1 = 0$. Note also that $\ln 2 > 0$, so the proportionality constant is less than 0, and the exponential function represents exponential decay.

(b)

$$Q(100) = 50e^{-\frac{\ln 2}{578}(100)}$$
$$\approx 44.35 \text{ mg}$$

(c)

$$10 = Q(t) = 50e^{-\frac{\ln 2}{578}t}$$

## 4.4. EXPONENTIAL GROWTH AND DECAY

$$e^{-\frac{\ln 2}{578}t} = \frac{1}{5}$$

$$-\frac{\ln 2}{578}t = \ln \frac{1}{5} = \ln 1 - \ln 5$$

$$= -\ln 5$$

$$t = \frac{578 \ln 5}{\ln 2} \approx 1342.1 \text{ hours} \approx 56 \text{ days}$$

(d)

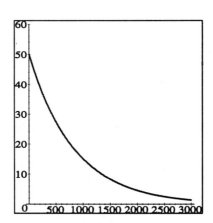

**Example 188** *(Exercise Set 4.4, Exercise 7)* Find the half-life of a radioactive substance that decays 3% in 5 years.

Solution: The quantity of the radioactive substance present at time $t$ is

$$Q(t) = Q_0 e^{kt}.$$

If the substance decays 3% in 5 years, then after 5 years there still remains 97% of the original amount. This gives the equation

$$0.97 Q_0 = Q(5) = Q_0 e^{5k}$$

from which the proportionality constant $k$ can be found. We do not need to know the actual initial amount of the substance. Solving gives

$$0.97 Q_0 = Q(5) = Q_0 e^{5k}$$
$$e^{5k} = 0.97$$
$$5k = \ln(0.97)$$
$$k = \frac{\ln(0.97)}{5}$$

and
$$Q(t) = Q_0 e^{\frac{\ln(0.97)}{5}t}.$$
To find the half-life, find $t$ so that
$$\frac{1}{2}Q_0 = Q_0 e^{\frac{\ln(0.97)}{5}t}$$
$$e^{\frac{\ln(0.97)}{5}t} = \frac{1}{2}$$
$$\frac{\ln(0.97)}{5}t = -\ln 2$$
$$t = -\frac{5\ln 2}{\ln(0.97)} \approx 113.8 \text{ years.}$$

Notice that the number of years is positive, since $\ln(0.97) < 0$.

**Example 189** *(Exercise Set 4.4, Exercise 9) The table gives estimates of the world population, in millions from 1950 to 1990, taken from the 1997 World Almanac.*

*(a) Use the exponential model and population figures from 1950 and 1960 to predict the world population in the years 2000 and 2050.*

*(b) Use the exponential model and population figures from 1980 and 1990 to predict the world population in the years 2000 and 2050.*

*(c) Use a graphing device to plot the original data points and the exponential models given in (a) and (b).*

| Year | 1950 | 1960 | 1970 | 1980 | 1990 |
|---|---|---|---|---|---|
| Population | 2513 | 3027 | 3678 | 4478 | 5321 |

Solution:

(a) The initial population is the 1950 statistic so the population at $t$ years after 1950 is
$$Q(t) = 2513 e^{kt}.$$
To find the proportionality constant we use the 1960 statistic. Since 1960 is 10 years after the initial date of 1950,
$$3027 = Q(10) = 2513 e^{10k}$$
$$e^{10k} = \frac{3027}{2513}$$

## 4.4. EXPONENTIAL GROWTH AND DECAY

$$10k = \ln \frac{3027}{2513}$$
$$k = \frac{1}{10} \ln \frac{3027}{2513}$$

and
$$Q(t) = 2513 e^{\frac{1}{10} \ln \frac{3027}{2513} t}.$$

Predictions:

Population in 2000: Since 2000 is 50 years after 1950, the population in 2000 is approximately

$$Q(50) = 2513 e^{\frac{1}{10} \ln \frac{3027}{2513} 50} \approx 6372 \text{ million}.$$

Population in 2050: Since 2050 is 100 years after 1950, the population in 2050 is approximately

$$Q(100) = 2513 e^{\frac{1}{10} \ln \frac{3027}{2513} 100} \approx 16,158 \text{ million}.$$

(b) The process is the same as in part (a) but replacing 1950 with 1980 and 1960 with 1990. So time is measured in years after 1980.

Initial Population: 4478, the 1980 population

Proportionality constant: 1990 is 10 years after 1980

$$5321 = Q(10) = 4478 e^{10k}$$
$$k = \frac{1}{10} \ln \frac{5321}{4478}$$

Population at time $t$:

$$Q(t) = 4478 e^{\frac{1}{10} \ln \frac{5321}{4478} t}$$

Population in 2000: 2000 is 20 years after 1980

$$Q(20) = 4478 e^{\frac{1}{10} \ln \frac{5321}{4478} (20)} \approx 6323 \text{ million}$$

Population in 2050: 2050 is 70 years after 1980

$$Q(70) = 4478 e^{\frac{1}{10} \ln \frac{5321}{4478} (70)} \approx 14,978 \text{ million}$$

(c) To plot the data points along with the exponential models on the same set of axes, it is necessary to shift the exponential functions so that time $t = 0$ corresponds with the initial time of year 1950 in the first model and 1980 in the second. So we plot

$$y = 2513 e^{\frac{1}{10} \ln \frac{3027}{2513} (t - 1950)} \quad \text{and} \quad y = 4478 e^{\frac{1}{10} \ln \frac{5321}{4478} (t - 1980)}.$$

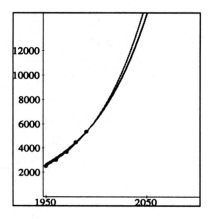

∎

# Chapter 5

# Conic Sections, Polar Coordinates, and Parametric Equations

## 5.1 Introduction

The graphs of the general *quadratic equation* in $x$ and $y$,

$$Ax^2 + Bxy + Cy^2 + Dx + Ey + F = 0,$$

are called *conic sections*. The three basic figures are the parabola, ellipse and hyperbola, although certain special, degenerate curves can also occur. When $B = 0$ and $AC = 0$ the graph is a parabola, when $B = 0$ and $AC > 0$ the graph is an ellipse, and when $B = 0$ and $AC < 0$ the curve is a hyperbola. When $B \neq 0$ the curve is a rotated conic in the plane.

Polar coordinates and parametric equations provide two additional methods for describing curves in the plane. They allow for the visualization of a greater variety of curves.

## 5.2 Parabolas

The graph of the familiar equation of the form $y = ax^2 + bx + c$ is a *parabola*, with axis parallel to the $y$-axis. A more general geometric definition of a parabola is the set of points equidistant from a given point, called the *focus*, and a given line, called the *directrix*. When the directrix is one of the coordinate axes the parabola is said to be in *standard form*. The *axis* of the parabola is the line through the focal point that is perpendicular to the directrix. The point of intersection of the axis and the parabola is the *vertex*. A useful tip to remember is a parabola *never* crosses its directrix.

### 5.2.1 Standard Position Parabolas

| Equation | Vertex | Focus | Directrix |
|---|---|---|---|
| $y = \frac{1}{4c}x^2$ | $(0,0)$ | $(0,c)$ | Horizontal: $y = -c$ |
| $x = \frac{1}{4c}y^2$ | $(0,0)$ | $(c,0)$ | Vertical: $x = -c$ |

Using the simple shifting properties, if the vertex is shifted to the point $(h, k)$, that is, $h$ units in the horizontal direction and $k$ units in the vertical direction, then the parabola has an equation of the form

$$(y-k) = \frac{1}{4c}(x-h)^2 \quad \text{or} \quad (x-h) = \frac{1}{4c}(y-k)^2.$$

**Example 190** *Find the vertex, directrix, and focus, and sketch the graph of the parabola.*
(a) $y = -\frac{1}{8}x^2$  (b) $4y^2 = 9x$

$-\frac{1}{8} = \frac{1}{4c} \qquad \frac{4c}{4} = \frac{-8}{4}, \quad c = -2$

Solution:
(a) The equation of the parabola is already in standard form, so we only need to determine the value of $c$. So we have

$$y = -\frac{1}{8}x^2 = \frac{1}{4c}x^2$$
$$\frac{1}{4c} = -\frac{1}{8}$$
$$4c = -8$$
$$c = -2.$$

Vertex: $(0,0)$, since the parabola is in standard form with vertex at the origin.
Focus: $(0, c) = (0, -2)$, since the axis of the parabola is along the $y$-axis.
Directrix: $y = -c = 2$
Since the parabola never crosses its directrix, the parabola opens downward as shown in the figure.

## 5.2. PARABOLAS

<u>Maximum or Minimum Value:</u> Since the curve is opening downward the vertex $(0,0)$ is a *maximum* point on the curve.
<u>Increasing:</u> The curve increases on the interval $(-\infty, 0)$.
<u>Decreasing:</u> The curve is decreasing on the interval $(0, \infty)$.

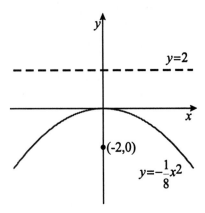

(b) A slight rewriting of the curve will put it in standard form.

$$4y^2 = 9x$$
$$x = \frac{4}{9}y^2$$
$$x = \frac{1}{9/4}y^2$$
$$4c = \frac{9}{4}$$
$$c = \frac{9}{16}$$

The parabola is in standard form with axis along the $x$-axis.
<u>Vertex:</u> $(0, 0)$
<u>Focus:</u> $(c, 0) = \left(\frac{9}{16}, 0\right)$
<u>Directrix:</u> $x = -c = -\frac{9}{16}$

Since the parabola never crosses its directrix, the parabola opens to the right, as shown in the figure. Notice also the equation does not define a function. This is seen from the figure, which shows that the graph does not satisfy the vertical line test. That is, for each $a > 0$ the vertical line $x = a$ crosses the curve in two points.

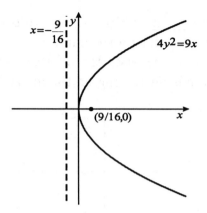

**Example 191** *Find the vertex, directrix and focus and sketch the graph of the parabola.*
  (a) (Exercise Set 5.2, Exercise 10)  $y^2 + 6y + 6 - 3x = 0$
  (b) (Exercise Set 5.2, Exercise 11)  $2x^2 + 4x - 9y + 20 = 0$

Solution:
(a) The first step is to group the $x$ and $y$ terms in order to rewrite the equation in standard form, with perhaps the vertex shifted. Since there are both $y$ and $y^2$ terms present, completing the square on these terms is necessary. Completing the square on the $y$ terms gives

$$y^2 + 6y = y^2 + 6y + \left(\frac{6}{2}\right)^2 - \left(\frac{6}{2}\right)^2$$
$$= y^2 + 6y + 9 - 9$$
$$= (y+3)^2 - 9.$$

Then

$$y^2 + 6y + 6 - 3x = 0$$
$$y^2 + 6y = 3x - 6$$
$$(y+3)^2 - 9 = 3x - 6$$
$$(y+3)^2 = 3x + 3$$
$$(y+3)^2 = 3(x+1)$$
$$(x+1) = \frac{1}{3}(y+3)^2.$$

## 5.2. PARABOLAS

The vertex of the parabola is $(-1,-3)$ and the graph of the parabola can be obtained from the graph of

$$x = \frac{1}{3}y^2,$$

which is a parabola in standard form with vertex at the origin and axis along the $x$-axis. Since

$$4c = 3, \quad c = \frac{3}{4}.$$

The focus, vertex, and directrix of the parabola $(x+1) = \frac{1}{3}(y+3)^2$ are obtained from shifting the focus, vertex, and directrix of the parabola $x = \frac{1}{3}y^2$. For example, the focus of $x = \frac{1}{3}y^2$ is $(c,0) = \left(\frac{3}{4},0\right)$, so the focus of $(x+1) = \frac{1}{3}(y+3)^2$ is

$$\left(\frac{3}{4}-1, 0-3\right) = \left(-\frac{1}{4},-3\right),$$

that is, shift the point $\left(\frac{3}{4},0\right)$ left 1 unit and 1 unit downward.

| Parabola | $x = \frac{1}{3}y^2$ | $(x+1) = \frac{1}{3}(y+3)^2$ |
|---|---|---|
| $c$ | $\frac{3}{4}$ | $\frac{3}{4}$ |
| Vertex | $(0,0)$ | $(-1,-3)$ |
| Focus | $(c,0) = \left(\frac{3}{4},0\right)$ | $\left(\frac{3}{4}-1, 0-3\right) = \left(-\frac{1}{4},-3\right)$ |
| Directrix | $x = -c = -\frac{3}{4}$ | $x = -c-1 = -\frac{3}{4}-1 = -\frac{7}{4}$ |

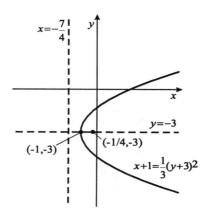

(b)

**Rewrite the equation:** Complete the square on the $x$ terms.

$$\begin{aligned} 2x^2 + 4x - 9y + 20 &= 0 \\ 2(x^2 + 2x) &= 9y - 20 \\ 2(x^2 + 2x + 1 - 1) &= 9y - 20 \\ 2(x+1)^2 - 2 &= 9y - 20 \\ 2(x+1)^2 &= 9y - 18 \\ 2(x+1)^2 &= 9(y-2) \\ (y-2) &= \frac{2}{9}(x+1)^2 \end{aligned}$$

The parabola is obtained by shifting, 1 unit to the left and 2 units upward, the parabola $y = \frac{2}{9}x^2$ that is in standard form with axis along the $y$-axis and

$$\begin{aligned} \frac{1}{4c} &= \frac{2}{9} \\ 4c &= \frac{9}{2} \\ c &= \frac{9}{8}. \end{aligned}$$

| Parabola | $y = \frac{2}{9}x^2$ | $(y-2) = \frac{2}{9}(x+1)^2$ |
|---|---|---|
| $c$ | $\frac{9}{8}$ | $\frac{9}{8}$ |
| Vertex | $(0,0)$ | $(-1, 2)$ |
| Focus | $(0, c) = \left(0, \frac{9}{8}\right)$ | $\left(-1, 2 + \frac{9}{8}\right) = \left(-1, \frac{25}{8}\right)$ |
| Directrix | $y = -c = -\frac{9}{8}$ | $y = -c + 2 = -\frac{9}{8} + 2 = \frac{7}{8}$ |

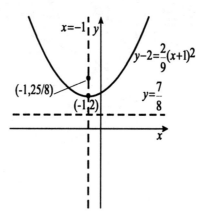

## 5.2. PARABOLAS

∎

**Example 192** *Determine the equation of the parabola that satisfies the given conditions.*
  *(a) (Exercise Set 5.2, Exercise 18)   Focus at $(-2,2)$, directrix $x = 2$.*
  *(b) Vertex at $(3,4)$,   Focus at $(3,6)$.*

Solution:
(a) The equation of a parabola requires the vertex and the value of c. The vertex of a parabola lies *midway* between the focus and the directrix on the line through the focus and perpendicular to the directrix. The horizontal distance between $(-2,2)$ and the vertical line $x = 2$ is 4, so the vertex is 2 units to the right of the focus $(-2,2)$, and hence is the point $(0,2)$.

If the focus had been $(-2,0)$ rather than $(-2,2)$, then the conditions would describe a parabola in standard position with vertex at the origin and equation
$$x = \frac{1}{4(-2)}y^2 = -\frac{1}{8}y^2.$$

The parabola described by the conditions is obtained by a vertical shift 2 units upward, of the standard position parabola. So
$$x = -\frac{1}{8}(y-2)^2.$$

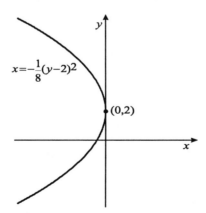

(b) Since the axis of the parabola is parallel to one of the coordinate axes and passes through the vertex $(3,4)$ and the focus $(3,6)$, the axis is vertical. The distance between the vertex and the focus is $6 - 4 = 2$ (both points are on the vertical line $x = 3$), and since the focus is above the vertex, the parabola opens upward, with $c = 2$. The directrix is 2 units below the vertex and has equation $y = 2$. The equation of the parabola is

$$y - 4 = \frac{1}{4(2)}(x - 3)^2$$

$$y - 4 = \frac{1}{8}(x - 3)^2.$$

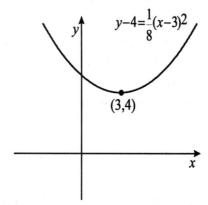

■

**Example 193** *Determine the equation of the parabola that satisfies the given conditions.*

(a) (Exercise Set 5.2, Exercise 22)  *Vertex at $(1,0)$; axis parallel to the $y$-axis; passing through the point $(5,6)$.*

(b) (Exercise Set 5.2, Exercise 26)  *Vertex at $(1,0)$; axis parallel to the $x$-axis; passing through the point $(5,6)$.*

Solution:
(a) The parabola in standard position with axis the $y$-axis has equation

$$y = \frac{1}{4c}x^2,$$

and the parabola with vertex at $(1,0)$ is given by

$$y = \frac{1}{4c}(x - 1)^2.$$

## 5.2. PARABOLAS

This uses the first two pieces of information. To use the third, observe that if the curve passes through $(5,6)$, then this point must satisfy the equation for the curve, so

$$6 = \frac{1}{4c}(5-1)^2$$
$$= \frac{1}{4c} \cdot 16$$
$$= \frac{4}{c}$$
$$c = \frac{4}{6} = \frac{2}{3}.$$

The equation of the parabola is then

$$y = \frac{1}{4\left(\frac{2}{3}\right)}(x-1)^2$$
$$y = \frac{3}{8}(x-1)^2.$$

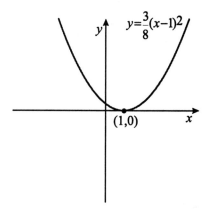

(b) The parabola in standard position with axis the $x$-axis has equation

$$x = \frac{1}{4c}y^2,$$

and the parabola with vertex at $(1,0)$ is given by

$$x - 1 = \frac{1}{4c}y^2.$$

The parabola passes through $(5,6)$, so

$$5 - 1 = \frac{1}{4c}(6)^2$$
$$4 = \frac{1}{4c} \cdot 36 = \frac{9}{c}$$
$$c = \frac{9}{4}.$$

The equation of the parabola is then

$$x - 1 = \frac{1}{4\left(\frac{9}{4}\right)}y^2$$
$$x - 1 = \frac{1}{9}y^2.$$

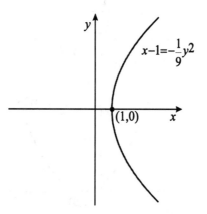

∎

## 5.2.2 Applications

**Example 194** *(Exercise Set 5.2, Exercise 30) A driving light has a parabolic cross section with a depth of 2 in. and a cross section height of 4 in. Where should the light source be placed to produce a parallel beam of light?*

Solution: Light rays emitted from the focus of a parabolic surface are reflected off the surface in parallel rays creating a concentrated beam of light. So the light source should be placed at the focus of the parabolic reflector. The information given does not allow us to find the location of the focus

## 5.2. PARABOLAS

directly, but we can find the general form for the equation of the parabola. This will allow us to determine the value of $c$, which will tell us where to place the light.

The parabolic cross section of the light is shown in the figure. Since it is a parabola in standard position with axis along the $x$-axis, the equation has the form

$$x = \frac{1}{4c}y^2.$$

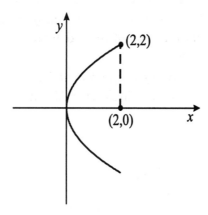

Since the parabola passes through the point $(2,2)$,

$$\begin{aligned} 2 &= \frac{1}{4c}(2)^2 \\ 2 &= \frac{1}{c} \\ c &= \frac{1}{2}. \end{aligned}$$

So the focus of the parabola is $\left(\frac{1}{2}, 0\right)$. To produce a parallel beam of light, the light source is placed at the focus, that is,

$$\frac{1}{2} \text{ inch from the vertex of the light.}$$

■

**Example 195** *(Exercise Set 5.2, Exercise 31) A ball thrown horizontally from the top edge of a building follows a parabolic curve with vertex at the top edge of the building and axis along the side of the building. The ball* passes through a point 100 ft from the building when it is a vertical distance of 16 ft from the top.

(a) How far from the building will the ball land if the building is 64 ft high?

(b) Suppose instead the ball is thrown from the top of the Sears Tower in Chicago, which has a height of 1450 ft. Recompute the answer to (a).

Solution: If we knew the equation of the parabolic path of the ball, then the solution to each part, is the value of $x$ when $y = 0$. The information gives one point on the parabola and the standard form for the parabola, which is enough to find the equation.

(a) Set up a $xy$-coordinate system so that the $x$-axis runs along the base of the building and the $y$-axis coincides with the edge of the building, as shown in the figure.

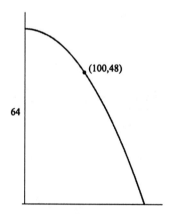

The path of the ball is a parabola in standard position with axis along the $y$-axis, opening downward, and whose vertex has been shifted to the point $(0, 64)$, since the building is 64 ft high. The equation of the parabola is

$$y - 64 = \frac{1}{4c}x^2.$$

Since the ball is 16 ft below the top of the building when it is 100 ft away from the edge, the parabola passes through the point

$$(100, 64 - 16) = (100, 48).$$

## 5.2. PARABOLAS

So

$$48 - 64 = \frac{100^2}{4c}$$
$$4c = -\frac{100^2}{16}$$
$$4c = -625$$

and the equation of the parabola is

$$y - 64 = -\frac{1}{625}x^2.$$

The ball hits the ground where the height $y$ is 0. Setting $y = 0$ and solving for $x$ gives

$$-64 = -\frac{1}{625}x^2$$
$$x^2 = 625 \cdot 64 = 40000$$
$$x = \sqrt{40000} = 200.$$

The ball hits the ground 200 ft from the building.

(b) The analysis is the same as that done in part (a), except now the building is 1450 ft high instead of 16 ft high. So,

$$y - 1450 = \frac{1}{4c}x^2$$

and the point

$$(100, 1450 - 16) = (100, 1434)$$

lies on the parabola. So

$$1434 - 1450 = \frac{100^2}{4c}$$
$$4c = -\frac{100^2}{16} = -625$$

which is the same as before. The equation of the parabola is

$$y - 1450 = -\frac{1}{625}x^2,$$

and when $y = 0$,

$$-1450 = -\frac{1}{625}x^2$$
$$x = \sqrt{1450 \cdot 625} \approx 952.$$

The ball hits the ground approximately 952 ft from the building. ∎

## 5.3 Ellipses

An *ellipse* is a set of points in a plane for which the sum of the distances from two fixed points is a given constant. The two fixed points are called the *focal points*, the line passing through the focal points is called the *axis*, and the points of intersection of the axis and the ellipse are called the *vertices*. An ellipse centered at the origin and with axis along one of the coordinate axes is said to be in *standard position*.

### 5.3.1 Standard Position Ellipses

| Equation: $a > b$ | $\frac{x^2}{a^2} + \frac{y^2}{b^2} = 1$ | $\frac{y^2}{a^2} + \frac{x^2}{b^2} = 1$ |
|---|---|---|
| Axis | $x$-axis | $y$-axis |
| Vertices | $(-a, 0), (a, 0)$ | $(0, -a), (0, a)$ |
| Focal Points: $c^2 = a^2 - b^2$ | $(-c, 0), (c, 0)$ | $(0, -c), (0, c)$ |
| Other Intercepts | $y: (0, -b), (0, b)$ | $x: (-b, 0), (b, 0)$ |
| Center | $(0, 0)$ | $(0, 0)$ |
| Eccentricity: $c = \sqrt{a^2 - b^2}$ | $e = \frac{c}{a}$ | $e = \frac{c}{a}$ |

The *center* of an ellipse in standard position is the origin, which is the midpoint of the line segment connecting the vertices. The axis of an ellipse is also called the *major axis* and the line segment connecting the other intercepts and perpendicular to the axis is called the *minor axis*.

If the center is shifted to the point $(h, k)$, then the ellipse will have an equation of the form

$$\frac{(x-h)^2}{a^2} + \frac{(y-k)^2}{b^2} = 1 \quad \text{or} \quad \frac{(y-k)^2}{a^2} + \frac{(x-h)^2}{b^2} = 1. \quad (5.1)$$

## 5.3. ELLIPSES

**Example 196** *Find the vertices and focal points and sketch the graph of the ellipse.*

(a) $\dfrac{x^2}{16} + \dfrac{y^2}{9} = 1$  (b) $16x^2 + 9y^2 = 144$

Solution:

(a) The equation is in standard form and, since the denominator of the $x^2$ term is larger than the denominator of the $y^2$ term, the axis of the ellipse is on the $x$-axis, with center at the origin. So

$$a^2 = 16, \quad a = 4$$
$$b^2 = 9, \quad b = 3$$
$$c = \sqrt{a^2 - b^2} = \sqrt{16 - 9} = \sqrt{7}.$$

<u>Vertices:</u> $(-a, 0) = (-4, 0), (a, 0) = (4, 0)$
<u>Focal Points:</u> $(-c, 0) = (-\sqrt{7}, 0), (c, 0) = (\sqrt{7}, 0)$
<u>$y$-intercepts:</u> $(0, -b) = (0, -3), (0, b) = (0, 3)$

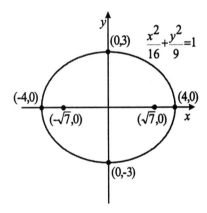

(b) First get the equation in standard form by dividing both sides of the equation by 144.

$$16x^2 + 9y^2 = 144$$
$$\dfrac{x^2}{9} + \dfrac{y^2}{16} = 1$$

Notice that the equation is similar to the equation in part (a), with the roles of $a$ and $b$ reversed. That is, the axis of the ellipse is the $y$-axis, although we still have $a = 4, b = 3$, and $c = \sqrt{7}$. So

Vertices: $(0, -a) = (0, -4), (0, a) = (0, 4)$
Focal Points: $(0, -c) = (0, -\sqrt{7}), (0, c) = (0, \sqrt{7})$
x-intercepts: $(-b, 0) = (-3, 0), (b, 0) = (3, 0)$
Notice in both parts (a) and (b) the length of the major axis is 8, and the length of the minor axis is 6.

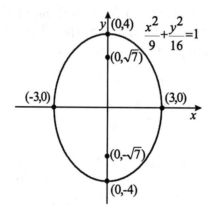

■

**Example 197** *(Exercise Set 5.3, Exercise 9)* Find the vertices and focal points and sketch the graph of $x^2 + 4y^2 - 2x - 16y + 13 = 0$.

Solution: The key to recognizing the ellipse is to group the $x$ terms and group the $y$ terms and complete the square on both. So

$$\begin{aligned} x^2 + 4y^2 - 2x - 16y + 13 &= 0 \\ x^2 - 2x + 4y^2 - 16y &= -13 \\ x^2 - 2x + 1 - 1 + 4(y^2 - 4y + 4 - 4) &= -13 \\ (x-1)^2 + 4(y-2)^2 &= -13 + 1 + 16 \\ (x-1)^2 + 4(y-2)^2 &= 4 \\ \frac{(x-1)^2}{4} + (y-2)^2 &= 1. \end{aligned}$$

The equation describes an ellipse that is obtained from an ellipse in standard form with the center shifted to the point $(1, 2)$. The ellipse in standard form is

$$\frac{x^2}{4} + y^2 = 1,$$

## 5.3. ELLIPSES

which has major axis on the $x$-axis, with $a = 2$, $b = 1$, center $(0, 0)$, vertices $(-2, 0)$ and $(2, 0)$. The focal points are $(-\sqrt{3}, 0)$ and $(\sqrt{3}, 0)$ since $c = \sqrt{a^2 - b^2} = \sqrt{4 - 1} = \sqrt{3}$.

The shifted ellipse satisfies:

Major axis: horizontal; on the line $y = 2$
Center: $(1, 2)$; shifted one unit right and 2 units upward from the origin
Vertices: $(-a + 1, 2) = (-1, 2), (a + 1, 2) = (3, 2)$
Focal points: $(-c + 1, 2) = (-\sqrt{3} + 1, 2), (c + 1, 2) = (\sqrt{3} + 1, 2)$
Minor axis: vertical; on the line $x = 1$
Minor axis intercepts: $(1, -b + 2) = (1, 1), (1, b + 2) = (1, 3)$

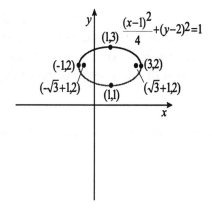

■

**Example 198** *Find an equation of the ellipse that satisfies the stated conditions.*

*(a) (Exercise Set 5.3, Exercise 11) The $x$-intercepts at $(\pm 4, 0)$ and $y$-intercepts at $(\pm 3, 0)$.*

*(b) (Exercise Set 5.3, Exercise 16) Foci at $(3, 0)$ and $(1, 0)$, a vertex at $(0, 0)$.*

Solution:

(a) Since the $x$- and $y$-intercepts are centered about the origin, the ellipse is in standard position. The distance between $(-4, 0)$ and $(4, 0)$ is 8 and the distance between $(-3, 0)$ and $(3, 0)$ is only 6, so the major axis is on the $x$-axis. Also,

$$a = 4, \quad b = 3$$

and the equation of the ellipse is

$$\frac{x^2}{16} + \frac{y^2}{9} = 1.$$

(b) Since the foci are on the $x$-axis, the major axis of the ellipse is also on the $x$-axis. To find $c$, find the midpoint of the line segment connecting the foci. Then $c$ will be the distance from the midpoint to either of the foci. The midpoint, which is also the center of the ellipse, occurs at

$$\left(\frac{3+1}{2}, 0\right) = (2, 0)$$
$$c = 3 - 2 = 1.$$

Since one vertex is $(0,0)$, which is 1 unit horizontally to the left of the focus $(1,0)$, the other vertex is 1 unit to the right of the focus $(3,0)$ and is $(4,0)$. The length of the major axis, which is the distance between the vertices, is 4. So $a$, which is half the length of the major axis is

$$a = 2.$$

Having both $a$ and $c$, we can now find $b$ and write the equation of the ellipse. So

$$c^2 = a^2 - b^2$$
$$b^2 = a^2 - c^2$$
$$b^2 = 4 - 1 = 3$$

and the equation of the ellipse is

$$\frac{(x-2)^2}{4} + \frac{y^2}{3} = 1.$$

■

## 5.4  Hyperbolas

A *hyperbola* is a set of points in the plane for which the magnitude of the difference between the distances from two fixed points is a given constant. The two fixed points are called the *foci*, the line passing through the focal points is called the *axis*, and the points of intersection of the axis with the hyperbola are called the *vertices*. A hyperbola centered at the origin and with axis along one of the coordinate axes is in *standard position*.

## 5.4. HYPERBOLAS

### 5.4.1 Standard Position Hyperbolas

| Equation | $\dfrac{x^2}{a^2} - \dfrac{y^2}{b^2} = 1$ | $\dfrac{y^2}{a^2} - \dfrac{x^2}{b^2} = 1$ |
| --- | --- | --- |
| Axis | $x$-axis | $y$-axis |
| Vertices | $(-a, 0), (a, 0)$ | $(0, -a), (0, a)$ |
| Focal points: $c^2 = a^2 + b^2$ | $(-c, 0), (c, 0)$ | $(0, -c), (0, c)$ |
| Center | $(0, 0)$ | $(0, 0)$ |
| Asymptotes | $y = \pm \dfrac{b}{a} x$ | $y = \pm \dfrac{a}{b} x$ |
| Eccentricity: $c = \sqrt{a^2 + b^2}$ | $e = \dfrac{c}{a}$ | $e = \dfrac{c}{a}$ |

If the center of the hyperbola in standard position is shifted to the point $(h, k)$, then the hyperbola will have an equation of the form

$$\frac{(x-h)^2}{a^2} - \frac{(y-k)^2}{b^2} = 1 \quad \text{or} \quad \frac{(y-k)^2}{a^2} - \frac{(x-h)^2}{b^2} = 1 \qquad (5.2)$$

and the asymptotes become

$$y - k = \pm \frac{b}{a}(x - h) \quad \text{or} \quad y - k = \pm \frac{a}{b}(x - h).$$

**Example 199** *Find the vertices, focal points, eccentricity, and equations of the asymptotes, and sketch the graph of the hyperbola.*
(a) $\dfrac{x^2}{9} - \dfrac{y^2}{16} = 1$  (b) $4y^2 - 25x^2 = 100$

Solution:
(a) The equation is in standard form. Since the $y^2$ term is negative the axis of the hyperbola is along the $x$-axis with center at the origin. So

$$\begin{aligned} a^2 &= 9, \quad a = 3 \\ b^2 &= 16, \quad b = 4 \\ c &= \sqrt{a^2 + b^2} = \sqrt{9 + 16} = 5. \end{aligned}$$

<u>Vertices:</u> $(-a, 0) = (-3, 0), (a, 0) = (3, 0)$
<u>Focal points:</u> $(-c, 0) = (-5, 0), (c, 0) = (5, 0)$
<u>Eccentricity:</u> $e = \dfrac{c}{a} = \dfrac{5}{3}$
<u>Asymptotes:</u> $y = \pm \dfrac{b}{a} x = \pm \dfrac{4}{3} x$

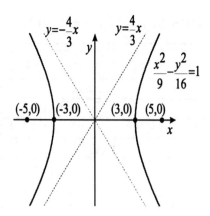

(b) First place the equation in standard form by dividing both sides of the equation by 100.

$$4y^2 - 25x^2 = 100$$
$$\frac{y^2}{25} - \frac{x^2}{4} = 1.$$

Since the $x^2$ term is negative, the axis is along the $y$-axis with center at the origin. So

$$a^2 = 25, a = 5$$
$$b^2 = 4, b = 2$$
$$c = \sqrt{a^2 + b^2} = \sqrt{29}.$$

<u>Vertices:</u> $(0, -a) = (0, -5), (0, a) = (0, 5)$
<u>Focal points:</u> $(0, -c) = (0, -\sqrt{29}), (0, c) = (0, \sqrt{29})$
<u>Eccentricity:</u> $e = \frac{c}{a} = \frac{\sqrt{29}}{5}$
<u>Asymptotes:</u> $y = \pm \frac{a}{b}x = \pm \frac{5}{2}x$

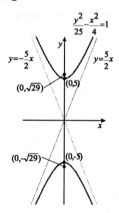

## 5.4. HYPERBOLAS

■

**Example 200** *(Exercise Set 5.4, Exercise 11) Find the vertices, focal points, eccentricity, and equations of the asymptotes, and sketch the graph of the hyperbola* $9x^2 - 4y^2 - 18x - 8y = 31$.

Solution: To compare the equation with the equation of a hyperbola in standard form, complete the square on both the $x$- and the $y$-terms. So,

$$\begin{aligned} 9x^2 - 4y^2 - 18x - 8y &= 31 \\ 9x^2 - 18x - 4y^2 - 8y &= 31 \\ 9(x^2 - 2x) - 4(y^2 + 2y) &= 31 \\ 9(x^2 - 2x + 1 - 1) - 4(y^2 + 2y + 1 - 1) &= 31 \\ 9(x-1)^2 - 9 - 4(y+1)^2 + 4 &= 31 \\ 9(x-1)^2 - 4(y+1)^2 &= 36 \\ \frac{(x-1)^2}{4} - \frac{(y+1)^2}{9} &= 1. \end{aligned}$$

The graph of the hyperbola is obtained from shifting the graph of the hyperbola in standard position, $\frac{x^2}{4} - \frac{y^2}{9} = 1$, to the right 1 unit and 1 unit downward. The standard position hyperbola has $a = 2, b = 3, c = \sqrt{13}$, vertices $(-2,0), (2,0)$, focal points $(-\sqrt{13},0), (\sqrt{13},0)$, and asymptotes $y = \pm\frac{3}{2}x$.
The hyperbola in this problem has the following properties.
Vertices: $(-a+1, -1) = (-1, -1), (a+1, -1) = (3, -1)$
Center: $(1, -1)$
Focal points: $(-c+1, -1) = (-\sqrt{13}+1, -1), (c+1, -1) = (\sqrt{13}+1, -1)$
Eccentricity: $e = \frac{c}{a} = \frac{\sqrt{13}}{2}$
Asymptotes:

$$\begin{aligned} y+1 &= \pm\frac{b}{a}(x-1) \\ y+1 &= \pm\frac{3}{2}(x-1) \\ y &= \frac{3}{2}x - \frac{5}{2}; y = -\frac{3}{2}x + \frac{1}{2} \end{aligned}$$

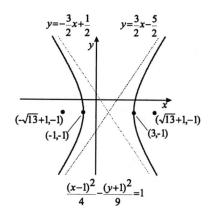

**Example 201** *Find an equation of the hyperbola that satisfies the stated conditions.*

(a) *(Exercise Set 5.4, Exercise 16)*   Foci at $(\pm 13, 0)$, vertices at $(\pm 5, 0)$.
(b) *(Exercise Set 5.4, Exercise 22)*   Foci at $(6, 1)$ and $(-2, 1)$, equations of asymptotes $y = \frac{3}{4}x - \frac{1}{2}$ and $y = -\frac{3}{4}x + \frac{5}{2}$.

Solution:
(a) Since the foci and vertices are on the $x$-axis and centered about the origin, the hyperbola is in standard form with axis on the $x$-axis and equation of the form

$$\frac{x^2}{a^2} - \frac{y^2}{b^2} = 1.$$

The vertices are $(\pm a, 0) = (\pm 5, 0)$, so $a = 5$. The foci are $(\pm c, 0) = (\pm 13, 0)$, so $c = 13$. To find the equation we need the value $b^2$, which we obtain from the equation

$$\begin{aligned} c^2 &= a^2 + b^2 \\ b^2 &= c^2 - a^2 \\ &= 169 - 25 \\ &= 144. \end{aligned}$$

The equation of the hyperbola is

$$\frac{x^2}{25} - \frac{y^2}{144} = 1.$$

## 5.4. HYPERBOLAS

(b) Since the foci both lie on the horizontal line $y = 1$, the axis of the hyperbola is parallel to the $x$-axis and the equation has the form

$$\frac{(x-h)^2}{a^2} - \frac{(y-k)^2}{b^2} = 1,$$

where $(h, k)$ is the center. The center is midway between the foci, so the center is the point

$$\left(\frac{6-2}{2}, \frac{1+1}{2}\right) = (2, 1),$$

and the equation can be written as

$$\frac{(x-2)^2}{a^2} - \frac{(y-1)^2}{b^2} = 1.$$

To find $a$ and $b$, use the information from the asymptotes. That is, the asymptotes are of the form

$$y - 2 = \pm\frac{b}{a}(x - 1),$$

so

$$\frac{b}{a} = \frac{3}{4}$$

$$b = \frac{3a}{4}.$$

Another equation is needed to find the two unknowns $a$ and $b$. But the value $c$ is half the distance between the foci, so

$$c = \frac{1}{2} \cdot 8 = 4$$

and

$$c^2 = a^2 + b^2$$

$$16 = a^2 + \left(\frac{3a}{4}\right)^2$$

$$16 = a^2 + \frac{9a^2}{16}$$

$$16 = \frac{25a^2}{16}$$

$$a^2 = \frac{256}{25}$$

$$b^2 = \frac{9a^2}{16} = \frac{9}{16} \cdot \frac{256}{25} = \frac{144}{25}.$$

The equation of the hyperbola is

$$\frac{25(x-2)^2}{256} - \frac{25(y-1)^2}{144} = 1.$$

∎

## 5.4.2 Applications

**Example 202** *(Exercise Set 5.4, Exercise 26)* *A company has two manufacturing plants that produce identical automobiles. Because of differing manufacturing and labor conditions in the plants, it costs $130 more to produce a car in plant A than in plant B. The shipping costs from both plants are the same, $1 per mile, as are the loading and unloading costs, $25 per car. State the criteria for determining from which plant a car should be shipped.*

Solution: Let the destination be denoted $C$. Then the total cost of production and delivery of an car from plants $A$ and $B$ to the destination $C$ are

$$\begin{aligned} T_A &= \$130 + \text{(production cost of a car at plant } B) \\ &\quad + \$1 \cdot d(A,C) + \text{(unloading costs)} \end{aligned}$$

and

$$\begin{aligned} T_B &= \text{(production cost of a car at plant } B) \\ &\quad + \$1 \cdot d(B,C) + \text{(unloading costs)}. \end{aligned}$$

If
 $T_A = T_B$, then the cost is the same from either plant,
 $T_A > T_B$, then the cost is greater if shipped from plant $A$,
 $T_A < T_B$, then the cost is greater if shipped from plant $B$.
Equivalently, if
 $T_A - T_B = 0$, then the cost is the same from either plant,
 $T_A - T_B > 0$, then the cost is greater if shipped from plant $A$,
 $T_A - T_B < 0$, then the cost is greater if shipped from plant $B$.
The advantage to considering the difference in the total costs is that

$$T_A - T_B = 130 + d(A,C) - d(B,C)$$

## 5.5. POLAR COORDINATES

and the equation shows that the distances of $C$ from the two plants are the only important parameters in the problem. The three cases are:

1. $\underline{T_A - T_B = 0}$: No difference whether the car is shipped from $A$ or $B$.

$$T_A - T_B = 130 + d(A, C) - d(B, C) = 0$$
$$d(A, C) - d(B, C) = -130$$
$$d(B, C) - d(A, C) = 130$$

2. $\underline{T_A - T_B > 0}$: Ship from $B$.

$$T_A - T_B = 130 + d(A, C) - d(B, C) > 0$$
$$d(A, C) - d(B, C) > -130$$
$$d(B, C) - d(A, C) < 130$$

3. $\underline{T_A - T_B < 0}$: Ship from $A$.

$$T_A - T_B = 130 + d(A, C) - d(B, C) < 0$$
$$d(A, C) - d(B, C) < -130$$
$$d(B, C) - d(A, C) > 130$$

In each case the dividing boundary is given by a hyperbola.

■

## 5.5 Polar Coordinates

In the rectangular coordinate system, a point in the plane is specified by an ordered pair $(x, y)$ that describes the location of the point using a rectangular grid. The first coordinate specifies the vertical distance to the $x$-axis along the line perpendicular to the $x$-axis. The second coordinate specifies the horizontal distance to the $y$-axis along the line perpendicular to the $y$-axis.

The *polar coordinate* system represents a point in the plane as an ordered pair $(r, \theta)$. The location of the point uses a distance, $r$, from the point to a fixed point called the *pole*, and the angle, $\theta$, made by the ray from the pole to the point and a fixed half ray extending from the pole. This fixed half ray is called the *polar axis*.

It is also convenient to allow negative entries for the first coordinate of a point given in polar coordinates. If $r > 0$, then

$$(r, \theta) \text{ and } (-r, \theta + \pi) \text{ represent the same point.}$$

That is, the point $(-r, \theta)$, is obtained by reflecting the point $(r, \theta)$ through the origin.

**Example 203** *Plot the point with the given polar coordinates. Then give two other pairs of polar coordinates that represent the point, one with $r > 0$ and one with $r < 0$.*
(a) $\left(2, \frac{\pi}{4}\right)$ (b) $\left(-3, \frac{\pi}{3}\right)$ (c) $\left(1, -\frac{\pi}{6}\right)$

Solution:

(a) Rotate the polar axis $\frac{\pi}{4}$ radians, or 45°, in the counterclockwise direction and place the point on the rotated ray 2 units from the pole.

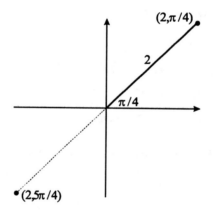

The polar representation of a point in the plane is *not* unique since every additional rotation of $2\pi$ radians, clockwise or counterclockwise, ends up at the same point. So another representation of the point is given by

$$\left(2, \frac{\pi}{4} + 2\pi\right) = \left(2, \frac{9\pi}{4}\right).$$

To find a representation of the point with $r < 0$, first find the point with $r > 0$, that is, the reflection through the origin of $\left(2, \frac{\pi}{4}\right)$. Then make the first coordinate negative. So

$$\left(2, \frac{\pi}{4} + \pi\right) = \left(2, \frac{5\pi}{4}\right) \text{ is the point opposite to } \left(2, \frac{\pi}{4}\right)$$

## 5.5. POLAR COORDINATES

and

$$\left(-2, \frac{5\pi}{4}\right) \text{ is opposite to } \left(2, \frac{5\pi}{4}\right) \text{ and so represents } \left(2, \frac{\pi}{4}\right).$$

(b) Possibilities are:
Second Representation, $r < 0$: $\left(-3, \frac{\pi}{3} + 2\pi\right) = \left(-3, \frac{7\pi}{3}\right)$
Third Representation, $r > 0$: $\left(3, \frac{\pi}{3} + \pi\right) = \left(3, \frac{4\pi}{3}\right)$

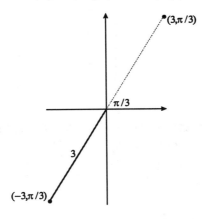

(c) The negative angle simply means rotate the polar axis in the clockwise direction to find the point. Subtracting $2\pi$ makes an additional rotation in the clockwise direction.
Second Representation, $r > 0$: $\left(1, -\frac{\pi}{6} - 2\pi\right) = \left(1, -\frac{13\pi}{6}\right)$
Third Representation, $r < 0$: $\left(-1, -\frac{\pi}{6} + \pi\right) = \left(-1, \frac{5\pi}{6}\right)$

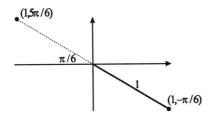

### 5.5.1 Polar Coordinates To Rectangular Coordinates

If the polar coordinates $(r, \theta)$ of a point in the plane are given, then the rectangular coordinates of the same point are

$$x = r \cos \theta \quad \text{and} \quad y = r \sin \theta.$$

These relationships are immediate consequences of the definitions of the cosine and sine of the angle $\theta$.

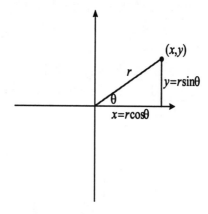

**Example 204** *Convert the polar coordinates to rectangular coordinates.*
(a) *(Exercise Set 5.5, Exercise 9)* $\left(1, \frac{\pi}{3}\right)$
(b) *(Exercise Set 5.5, Exercise 14)* $\left(-3, -\frac{7\pi}{6}\right)$

Solution:
(a) Since $r = 1$ and $\theta = \frac{\pi}{3}$,

$$\begin{aligned} x &= r\cos\theta = 1\cos\frac{\pi}{3} = \frac{1}{2} \\ y &= r\sin\theta = 1\sin\frac{\pi}{3} = \frac{\sqrt{3}}{2}. \end{aligned}$$

(b) Since the reference angle for $-\frac{7\pi}{6}$ is $\frac{\pi}{6}$ and $-\frac{7\pi}{6}$ is in quadrant II,

$$\cos\left(-\frac{7\pi}{6}\right) = -\cos\frac{\pi}{6} \quad \text{and} \quad \sin\left(-\frac{7\pi}{6}\right) = \sin\frac{\pi}{6}.$$

So

$$\begin{aligned} x &= -3\cos\left(-\frac{7\pi}{6}\right) = -3\left(-\cos\frac{\pi}{6}\right) = -3\left(-\frac{\sqrt{3}}{2}\right) = \frac{3\sqrt{3}}{2} \\ y &= -3\sin\left(-\frac{7\pi}{6}\right) = -3\sin\frac{\pi}{6} = -3\cdot\frac{1}{2} = -\frac{3}{2}. \end{aligned}$$

■

## 5.5. POLAR COORDINATES

**Example 205** *Convert the polar equation to a rectangular equation.*
  (a) *(Exercise Set 5.5, Exercise 23)* $r = 2\cos\theta$
  (b) $r^2 = \cos 2\theta$

Solution: The terms to look for are $x = r\cos\theta$, $x = r\sin\theta$ and $r^2 = x^2 + y^2$. This last relationship is just the Pythagorean Theorem. If these terms are not present

multiply both sides of the equation by $r$,

then make the replacements. So

$$\begin{aligned} r &= 2\cos\theta \\ r^2 &= 2r\cos\theta \\ x^2 + y^2 &= 2x \\ x^2 - 2x + y^2 &= 0. \end{aligned}$$

We recognize this as the equation of circle, but it is not in standard form. Complete the square on the $x$ term. Then

$$\begin{aligned} x^2 - 2x + y^2 &= 0 \\ x^2 - 2x + 1 - 1 + y^2 &= 0 \\ (x-1)^2 + y^2 &= 1 \end{aligned}$$

and this is the equation of the circle with radius 1 and center at the point $(1, 0)$.

(b) We can immediately make the substitution $r^2 = x^2 + y^2$ and get

$$\begin{aligned} r^2 &= \cos 2\theta \\ x^2 + y^2 &= \cos 2\theta. \end{aligned}$$

In the current form there is no way to replace the $\cos 2\theta$ term, but there is a trigonometric identity that is useful. If we rewrite the $\cos 2\theta$ in terms of just cosines and sines then we can use the substitutions $x = r\cos\theta$ and $y = r\sin\theta$. The identity we need is

$$\cos 2\theta = (\cos\theta)^2 - (\sin\theta)^2.$$

Then

$$\begin{aligned} r^2 &= \cos 2\theta \\ x^2 + y^2 &= \cos 2\theta \\ x^2 + y^2 &= (\cos\theta)^2 - (\sin\theta)^2 \\ &= \left(\frac{x}{r}\right)^2 - \left(\frac{y}{r}\right)^2 \\ x^2 + y^2 &= \frac{x^2}{r^2} - \frac{y^2}{r^2} \\ &= \frac{x^2}{x^2+y^2} - \frac{y^2}{x^2+y^2} \\ \left(x^2+y^2\right)^2 &= x^2 - y^2. \end{aligned}$$

■

## 5.5.2 Rectangular Coordinates To Polar Coordinates

If the rectangular coordinates $(x, y)$ of a point in the plane are given, then one set polar coordinates of the same point is obtained from

$$\tan\theta = \frac{y}{x}, \quad x \neq 0, \quad r^2 = x^2 + y^2,$$

where the sign of $r$ is chosen to ensure that the point is in the correct quadrant. When $x = 0$, we have $\theta = \frac{\pi}{2}$ and $r = y$.

**Example 206** *Convert the rectangular coordinates to polar coordinates.*
  (a) (Exercise Set 5.5, Exercise 19)   $(-2, 2)$
  (b) (Exercise Set 5.5, Exercise 20)   $(3\sqrt{3}, 3)$

Solution:
(a) Since $x = -2$ and $y = 2$,

$$\tan\theta = \frac{2}{-2} = -1$$

and so,

$$\theta = \frac{3\pi}{4} \text{ or } \theta = -\frac{\pi}{4}.$$

## 5.5. POLAR COORDINATES

To find $r$,

$$\begin{aligned} r^2 &= x^2 + y^2 \\ &= (-2)^2 + (2)^2 = 8 \\ r &= \pm\sqrt{8} = \pm 2\sqrt{2}. \end{aligned}$$

The point $(-2, 2)$ lies in quadrant II, so two different representations can be given as

$$\left(2\sqrt{2}, \frac{3\pi}{4}\right) \quad \text{and} \quad \left(-2\sqrt{2}, -\frac{\pi}{4}\right).$$

(b) Since

$$\tan\theta = \frac{3}{3\sqrt{3}} = \frac{1}{\sqrt{3}} = \frac{\sqrt{3}}{3}$$

we have

$$\theta = \frac{\pi}{6}, \frac{7\pi}{6}$$

and

$$\begin{aligned} r^2 &= \left(3\sqrt{3}\right)^2 + 3^2 \\ &= 36 \end{aligned}$$

so

$$r = \pm 6.$$

Since the point $(3\sqrt{3}, 3)$ lies in quadrant I, two possible polar representations are

$$\left(6, \frac{\pi}{6}\right) \quad \text{and} \quad \left(-6, \frac{7\pi}{6}\right).$$

■

**Example 207** *Convert the rectangular equation to polar coordinates.*
(a) (Exercise Set 5.5, Exercise 28)  $x^2 + y^2 = 2y$
(b) (Exercise Set 5.5, Exercise 29)  $y = x^2$

Solution:
(a) Using the relations $r^2 = x^2 + y^2$ and $y = r\sin\theta$,

$$\begin{aligned} x^2 + y^2 &= 2y \\ r^2 &= 2r\sin\theta \\ r &= 2\sin\theta. \end{aligned}$$

(b) Using the relations $x = r\cos\theta$ and $y = r\sin\theta$,

$$\begin{aligned} y &= x^2 \\ r\sin\theta &= (r\cos\theta)^2 \\ &= r^2(\cos\theta)^2 \end{aligned}$$

and

$$\begin{aligned} r &= \frac{\sin\theta}{(\cos\theta)^2} \\ &= \frac{\sin\theta}{\cos\theta} \cdot \frac{1}{\cos\theta} \end{aligned}$$

so

$$r = \tan\theta \sec\theta$$

## 5.5.3 Graphs of Polar Equations

The graph of a polar equation $r = f(\theta)$ is the collection of all points that have at least one polar representation that satisfies the equation.

| | |
|---|---|
| Circles | $r = a, r = a\cos\theta, r = a\sin\theta$ |
| Lines | $\theta = a$ |
| Cardioid | $r = a + a\cos\theta, r = a + a\sin\theta$ |
| Limacon without a loop: $a > b$ | $r = a + b\cos\theta, r = a + b\sin\theta$ |
| Limacon with a loop: $a < b$ | $r = a + b\cos\theta, r = a + b\sin\theta$ |
| Lemniscates | $r^2 = a^2\cos 2\theta, r^2 = a^2\sin 2\theta$ |
| Roses | $r = a\cos n\theta, r = a\sin n\theta$, leafs=$\begin{cases} 2n, & n \text{ even} \\ n, & n \text{ odd} \end{cases}$ |

**Example 208** *Sketch the graph of the polar equation.*
(a) *(Exercise Set 5.5, Exercise 35)* $r = 3\cos\theta$
(b) *(Exercise Set 5.5, Exercise 40)* $r = 1 + \sin\theta$

## 5.5. POLAR COORDINATES

**Solution:**
(a) Since the period of the cosine function is $2\pi$, the entire curve will be traced if $\theta$ varies between 0 and $2\pi$.

| $\theta$ | 0 | $\frac{\pi}{6}$ | $\frac{\pi}{4}$ | $\frac{\pi}{3}$ | $\frac{\pi}{2}$ | $\frac{2\pi}{3}$ | $\frac{3\pi}{4}$ | $\frac{5\pi}{6}$ | $\pi$ |
|---|---|---|---|---|---|---|---|---|---|
| $\cos\theta$ | 1 | $\frac{\sqrt{3}}{2}$ | $\frac{\sqrt{2}}{2}$ | $\frac{1}{2}$ | 0 | $-\frac{1}{2}$ | $-\frac{\sqrt{2}}{2}$ | $-\frac{\sqrt{3}}{2}$ | $-1$ |
| $r = 3\cos\theta$ | 3 | $\frac{3\sqrt{3}}{2}$ | $\frac{3\sqrt{2}}{2}$ | $\frac{3}{2}$ | 0 | $-\frac{3}{2}$ | $-\frac{3\sqrt{2}}{2}$ | $-\frac{3\sqrt{3}}{2}$ | $-3$ |
| $\theta$ | $\frac{7\pi}{6}$ | $\frac{5\pi}{4}$ | $\frac{4\pi}{3}$ | $\frac{3\pi}{2}$ | $\frac{5\pi}{3}$ | $\frac{7\pi}{4}$ | $\frac{11\pi}{6}$ | $2\pi$ | |
| $\cos\theta$ | $-\frac{\sqrt{3}}{2}$ | $-\frac{\sqrt{2}}{2}$ | $-\frac{1}{2}$ | 0 | $\frac{1}{2}$ | $\frac{\sqrt{2}}{2}$ | $\frac{\sqrt{3}}{2}$ | 1 | |
| $r = 3\cos\theta$ | $-\frac{3\sqrt{3}}{2}$ | $-\frac{3\sqrt{2}}{2}$ | $-\frac{3}{2}$ | 0 | $\frac{3}{2}$ | $\frac{3\sqrt{2}}{2}$ | $\frac{3\sqrt{3}}{2}$ | 3 | |

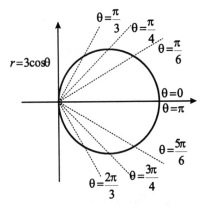

Notice how the points with $r < 0$ are reflected through the origin before they are plotted. Also, the graph is symmetric with respect to the polar axis, and in this example it would have been sufficient to only plot $\theta$ ranging from 0 to $\pi$. The curve appears to be the circle with center $\left(\frac{3}{2}, 0\right)$ and radius $\frac{3}{2}$. To see this algebraically, we convert the polar equation to a rectangular equation. That is,

$$r = 3\cos\theta$$
$$r^2 = 3r\cos\theta$$
$$x^2 + y^2 = 3x$$
$$x^2 - 3x + y^2 = 0$$
$$x^2 - 3x + \frac{9}{4} + y^2 = \frac{9}{4}$$
$$\left(x - \frac{3}{2}\right)^2 + y^2 = \left(\frac{3}{2}\right)^2.$$

(b) We could plot a number of points as in part (a), but another way to construct the graph is to use the values on the sine curve for $0 \leq \theta \leq 2\pi$.

| $\theta$ | $\sin\theta$ | $1+\sin\theta$ |
|---|---|---|
| Increases from 0 to $\frac{\pi}{2}$ | Increases from 0 to 1 | Increases from 1 to 2 |
| Increases from 0 to $\pi$ | Decreases from 1 to 0 | Decreases from 2 to 1 |
| Increases from $\pi$ to $\frac{3\pi}{2}$ | Decreases from 0 to $-1$ | Decreases from 1 to 0 |
| Increases from $\frac{3\pi}{2}$ to $2\pi$ | Increases from $-1$ to 0 | Increases from 0 to 1 |

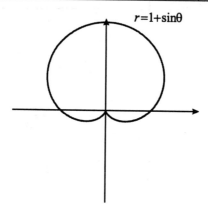

■

## 5.5.4 Intersection of Polar Curves

**Example 209** *(Exercise Set 5.5, Exercise 63) Find all points of intersection of the curves $r = 2 + 2\cos\theta$ and $r = -2\cos\theta$.*

Solution: We first take an algebraic approach and find all values of $\theta$ for which the two equations yield the same values of $r$. So we equate the two expressions for $r$ and solve for $\theta$ to give

$$\begin{aligned} 2 + 2\cos\theta &= -2\cos\theta \\ 4\cos\theta &= -2 \\ \cos\theta &= -\frac{1}{2} \\ \theta &= \frac{2\pi}{3} \quad \text{or} \quad \theta = \frac{4\pi}{3}. \end{aligned}$$

Do we have all the values of $\theta$? The answer is we may or we may not. The figure shows that the curves also intersect at the pole! The algebra did not yield this point.

## 5.6. CONIC SECTIONS IN POLAR COORDINATES

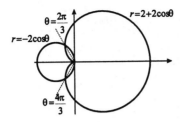

The reason is that the expressions $2+2\cos\theta$ and $-2\cos\theta$ are 0 for different values of $\theta$, so these values will not be found by equating the expressions. In this case, when

$$r = 2 + 2\cos\theta = 0$$
$$\cos\theta = -1$$
$$\theta = \pi$$

and when

$$r = -2\cos\theta = 0$$
$$\cos\theta = 0$$
$$\theta = \frac{\pi}{2}.$$

■

## 5.6 Conic Sections in Polar Coordinates

The graph of a polar equation of the form

$$r = \frac{ed}{1 \pm e\cos\theta} \quad \text{or} \quad r = \frac{ed}{1 \pm e\sin\theta}$$

is a conic section with eccentricity $e$. If

$$e = 1, \text{ the graph is a parabola}$$
$$e < 1, \text{ the graph is an ellipse}$$
$$e > 1, \text{ the graph is a hyperbola.}$$

The focus is at the pole and the directrix is $d$ units from the pole.

| Equation | Directrix |
|---|---|
| $r = \dfrac{ed}{1 + e\cos\theta}$ | Vertical, right of pole |
| $r = \dfrac{ed}{1 - e\cos\theta}$ | Vertical, left of pole |
| $r = \dfrac{ed}{1 + e\sin\theta}$ | Horizontal, above the pole |
| $r = \dfrac{ed}{1 - e\sin\theta}$ | Horizontal, below the pole |

**Example 210** *Sketch the graph of the conic section and find a corresponding rectangular equation.*

(a) (Exercise Set 5.6, Exercise 1) $\quad r = \dfrac{2}{1 + \cos\theta}$

(b) (Exercise Set 5.6, Exercise 5) $\quad r = \dfrac{3}{3 - \sin\theta}$

Solution:

(a) First determine $e$ and $d$. The polar equation has the form

$$r = \frac{ed}{1 + e\cos\theta}$$

with

$$\begin{aligned} e &= 1 \\ ed &= 2 \Rightarrow d = 2. \end{aligned}$$

Since $e = 1$, the conic is a parabola, and the directrix is vertical and 2 units to the right of the pole. The directrix has rectangular equation $x = 2$. Since the vertex is midway between the directrix and the pole, the vertex is at the point $(1, 0)$ (in rectangular or polar coordinates). The parabola can not cross the directrix and hence opens to the left. When

$$\begin{aligned} \theta &= \frac{\pi}{2}, \quad \cos\theta = 0 \quad \text{and} \quad r = 2 \\ \theta &= \frac{3\pi}{2}, \quad \cos\theta = 0 \quad \text{and} \quad r = 2. \end{aligned}$$

The graph is shown in the figure.

## 5.6. CONIC SECTIONS IN POLAR COORDINATES

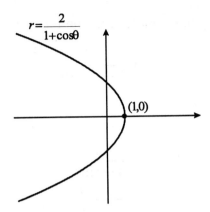

The rectangular equation of the parabola has the form

$$x - h = \frac{1}{4c}(y - k)^2.$$

To determine the rectangular equation, convert the polar equation to the corresponding rectangular form using the relations $x = r\cos\theta$, $y = r\sin\theta$, and $r^2 = x^2 + y^2$. Then

$$\begin{aligned}
r &= \frac{2}{1 + \cos\theta} \\
r + r\cos\theta &= 2 \\
r &= 2 - r\cos\theta = 2 - x \\
r^2 &= (x - 2)^2 \\
x^2 + y^2 &= x^2 - 4x + 4 \\
y^2 &= -4x + 4 \\
y^2 &= -4(x - 1) \\
x - 1 &= -\frac{1}{4}y^2.
\end{aligned}$$

In rectangular form the vertex of the parabola is $(1, 0)$ and $c = -1$. The parabola is obtained from the parabola in standard form

$$x = -\frac{1}{4}y^2$$

shifted 1 unit to the right. The focus is then $(c + 1, 0) = (-1 + 1, 0) = (0, 0)$, and the directrix is $x = 2$.

(b) First write the polar equation in the form

$$r = \frac{ed}{1 - e\sin\theta}.$$

Then,

$$r = \frac{3}{3 - \sin\theta}$$
$$= \frac{3}{3(1 - \frac{1}{3}\sin\theta)}$$

so

$$r = \frac{1}{1 - \frac{1}{3}\sin\theta}.$$

Since $e = \frac{1}{3}$, the conic is an ellipse and

$$ed = 1 \Rightarrow d = 3.$$

The conic is an ellipse with directrix horizontal and 3 units below the pole. The graph passes through the points

$$(1, 0), \quad (1, \pi), \quad \left(\frac{3}{2}, \frac{\pi}{2}\right), \quad \text{and} \quad \left(\frac{3}{4}, \frac{3\pi}{2}\right).$$

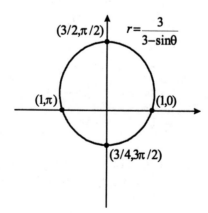

Rectangular equation:

$$r = \frac{3}{3 - \sin\theta}$$

## 5.6. CONIC SECTIONS IN POLAR COORDINATES

$$3r - r\sin\theta = 3$$
$$3r = 3 + r\sin\theta$$
$$3r = 3 + y$$
$$9r^2 = (3+y)^2$$
$$9x^2 + 9y^2 = 9 + 6y + y^2$$
$$9x^2 + 8y^2 - 6y = 9$$
$$9x^2 + 8\left(y^2 - \frac{3}{4}y\right) = 9$$
$$9x^2 + 8\left(y^2 - \frac{3}{4}y + \frac{9}{64} - \frac{9}{64}\right) = 9$$
$$9x^2 + 8\left(y - \frac{3}{8}\right)^2 = 9 + \frac{9}{8}$$
$$9x^2 + 8\left(y - \frac{3}{8}\right)^2 = \frac{81}{8}$$

■

### 5.6.1 Application

**Example 211** *(Exercise Set 5.6, Exercise 14) The earth moves in an elliptical orbit with the sun at one focal point and an eccentricity $e = 0.0167$. The major axis of the elliptical orbit is approximately $2.99 \times 10^8$ kilometers. Write a polar equation for this ellipse, assuming that the pole is at the sun.*

Solution: The elliptic orbit of the earth about the sun has an equation of the form
$$r = \frac{ed}{1 + e\cos\theta},$$
where the earth is closest the sun when $\theta = 0$ and furthest from the sun when $\theta = \pi$. The length of the major axis is

$$\frac{ed}{1 + e\cos\pi} - \frac{ed}{1 + e\cos 0} = \frac{ed}{1-e} - \frac{ed}{1+e}$$
$$= \frac{ed(1+e) - ed(1-e)}{(1-e)(1+e)}$$
$$= \frac{ed + e^2d - ed + e^2d}{1 - e^2}$$

$$= \frac{2e^2 d}{1-e^2}.$$

If the length of the major axis is approximately $2.99 \times 10^8$ and $e = 0.0167$, then

$$\begin{aligned} 2.99 \times 10^8 &= \frac{2e^2 d}{1-e^2} \\ d &= \frac{(2.99 \times 10^8)(1-e^2)}{2e^2} \\ &= \frac{(2.99 \times 10^8)(1-0.0167^2)}{2(0.0167)^2} \\ &\approx 5.4 \times 10^{11}. \end{aligned}$$

The polar equation of the earth's orbit about the sun is

$$r = \frac{0.0167(5.4 \times 10^{11})}{1 + 0.0167 \cos \theta} = \frac{9 \times 10^9}{1 + 0.0167 \cos \theta}.$$

■

## 5.7 Parametric Equations

In this method of representing curves in the plane, the $x$ and $y$ coordinates of a point on the curve are each specified separately by functions of a third variable, called the *parameter*. If the functions are $f$ and $g$ and the parameter $t$, then

$$x = f(t) \quad \text{and} \quad y = g(t)$$

are called the *parametric equations* for the curve. As $t$ varies, the point $(x, y) = (f(t), g(t))$ traces out the curve.

**Example 212** *Sketch the graph of the curve described by the parametric equations*

$$x = 2t - 1 \quad \text{and} \quad y = t^2 + 1$$

*and find a corresponding rectangular equation.*

Solution: First plot some representative points on the curve by making a table of values.

## 5.7. PARAMETRIC EQUATIONS

| $t$ | $x = 2t - 1$ | $y = t^2 + 1$ |
|---|---|---|
| 0 | −1 | 1 |
| 1 | 1 | 2 |
| 2 | 3 | 5 |
| −1 | −3 | 2 |
| −2 | −5 | 5 |

The points are plotted in the figure and a curve traced between them. The curve appears to be a parabola.

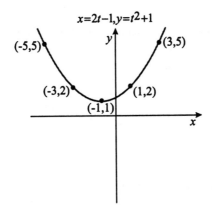

To find a corresponding rectangular equation for the curve, *eliminate the parameter*. That is, solve for $t$ in either the equation for $x$ or the equation for $y$, which ever is easiest, and then substitute the value for $t$ in the other equation. So

$$x = 2t - 1$$
$$2t = x + 1$$
$$t = \frac{x+1}{2}$$

and then substituting in the equation for $y$ gives

$$y = t^2 + 1$$
$$= \left(\frac{x+1}{2}\right)^2 + 1$$
$$= \frac{1}{4}(x+1)^2 + 1.$$

The equation describes a parabola with vertex at $(-1, 1)$ and opening upward.

∎

**Example 213** *Eliminate the parameter to find a corresponding rectangular equation for the curve and sketch.*
   (a) (Exercise Set 5.7, Exercise 3) $x = \sqrt{t}, y = t + 1$
   (b) $x = e^t - 2, y = e^{2t} + 3$
   (c) $x = 4 + 2\cos t, y = 6 + 2\sin t$

Solution:
(a) Solve for $t$ in the equation for $x$ and substitute in the equation for $y$. So

$$x = \sqrt{t}$$
$$(x)^2 = \left(\sqrt{t}\right)^2$$
$$t = x^2$$
$$y = x^2 + 1.$$

Be careful here that you have the *correct* curve. Is the curve the entire parabola given by the equation $y = x^2 + 1$? In this case the answer is no! The reason is that

$$\text{since } x = \sqrt{t}, \ x \text{ must be greater than or equal to } 0.$$

If

$$x \geq 0, \text{ then } y = x^2 + 1 \geq 1$$

and the curve consists *only* of the portion of $y = x^2 + 1$ to the right of the $y$-axis.

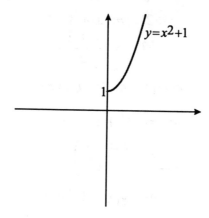

## 5.7. PARAMETRIC EQUATIONS

(b) In this example we do not have to find $t$ since finding an expression for $e^t$ will do. So

$$\begin{aligned} x &= e^t - 2 \\ e^t &= x + 2 \\ e^{2t} &= \left(e^t\right)^2 = (x+2)^2 \end{aligned}$$

and

$$y = e^{2t} + 3 = (x+2)^2 + 3$$

with the restriction that, since $e^t > 0$, we have $x = e^t - 2 > -2$.

The curve is then,

$$y = (x+2)^2 + 3, \text{ for } x > -2.$$

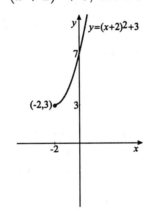

(c) In problems involving sine and cosine always be on the look out for

$$(\sin t)^2 + (\cos t)^2 = 1.$$

If $x - 4$ and $y - 6$ are each squared and then added, the resulting expression involves this basic identity and can then be simplified. That is,

$$\begin{aligned} (x-4)^2 + (y-6)^2 &= (4 + 2\cos t - 4)^2 + (6 + 2\sin t - 6)^2 \\ &= (2\cos t)^2 + (2\sin t)^2 \\ &= 4(\cos t)^2 + 4(\sin t)^2 \\ &= 4((\cos t)^2 + (\sin t)^2) \\ &= 4 \end{aligned}$$

which is the equation of circle with center $(4, 6)$ and radius 2.

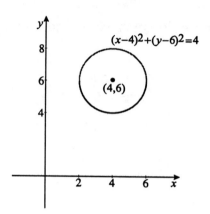

■

**Example 214** *(Exercise Set 5.7, Exercise 24) Find parametric equations for the line passing through the points $(5,3)$ and $(-2,7)$.*

Solution: First find the rectangular equation of the line. We need the slope, which is
$$m = \frac{7-3}{-2-5} = -\frac{4}{7}.$$
Using the point $(5,3)$ and the point-slope equation of a line gives
$$y - 3 = -\frac{4}{7}(x-5).$$
One set of parametric equations for the line can be found by setting
$$t = x - 5.$$
So
$$\begin{aligned} x &= t+5 \\ y - 3 &= -\frac{4}{7}(t+5-5) \\ y &= -\frac{4}{7}t + 3. \end{aligned}$$
A set of parametric equations is given by
$$x = t+5 \quad \text{and} \quad y = -\frac{4}{7}t + 3.$$

■

## 5.7. PARAMETRIC EQUATIONS

**Example 215** *(Exercise Set 5.7, Exercise 26) Find parametric equations in the parameter t, where $0 \le t \le 2\pi$, for the circle $x^2 + y^2 = r^2$ so that it is traced*

*(a) once around, counterclockwise, starting at $(r, 0)$*
*(b) twice around, counterclockwise, starting at $(r, 0)$*
*(c) three times around, counterclockwise, starting at $(-r, 0)$*
*(d) twice around, clockwise, starting at $(0, r)$*
*(e) three times around, clockwise, starting at $(0, r)$*

Solution: The idea is to recognize that if

$$x = r\cos t \quad \text{and} \quad y = r\sin t,$$

then

$$\begin{aligned} x^2 + y^2 &= (r\cos t)^2 + (r\sin t)^2 \\ &= r^2((\cos t)^2 + (\sin t)^2) \\ &= r^2, \end{aligned}$$

which is the equation of the circle. By slightly modifying these parametric equations the manner in which the circle is traced can be changed.

(a) No modification is required. The parametric equations

$$x = r\cos t \quad \text{and} \quad y = r\sin t, \quad \text{for} \quad 0 \le t \le 2\pi$$

trace the circle starting at $(r, 0)$ when $t = 0$, and ending back at $(r, 0)$ when $t = 2\pi$. Notice also that when

$$\begin{aligned} t &= \frac{\pi}{2}, & (x, y) &= (0, r) \\ t &= \pi, & (x, y) &= (-r, 0) \\ t &= \frac{3\pi}{2}, & (x, y) &= (0, -r). \end{aligned}$$

(b) To trace the circle twice, cut the period of the cosine and sine functions in half so that as $t$ varies from 0 to $2\pi$, the points on the circle are traced twice . To halve the period multiply the argument $t$, by 2. Note the period of $y = \cos 2t$ or if $y = \sin 2t$ is $\frac{2\pi}{2} = \pi$. So the parametric equations are

$$x = r\cos 2t \quad \text{and} \quad y = r\sin 2t, \quad \text{for } 0 \le t \le 2\pi.$$

Notice also that when

$$\begin{aligned} t &= 0, & (x,y) &= (r,0) \\ t &= \frac{\pi}{4}, & (x,y) &= (0,r) \\ t &= \frac{\pi}{2}, & (x,y) &= (-r,0) \\ t &= \frac{3\pi}{4}, & (x,y) &= (0,-r) \\ t &= \pi, & (x,y) &= (r,0). \end{aligned}$$

(c)
$$x = -r\cos 3t \quad \text{and} \quad y = -r\sin 3t, \quad \text{for } 0 \le t \le 2\pi$$

$$\begin{aligned} t &= 0, & (x,y) &= (-r,0) \\ t &= \frac{\pi}{6}, & (x,y) &= (0,-r) \\ t &= \frac{\pi}{3}, & (x,y) &= (r,0) \\ t &= \frac{\pi}{2}, & (x,y) &= (0,r) \\ t &= \frac{2\pi}{3}, & (x,y) &= (-r,0) \end{aligned}$$

(d)
$$x = r\sin 2t \quad \text{and} \quad y = r\cos 2t, \quad \text{for } 0 \le t \le 2\pi$$

$$\begin{aligned} t &= 0, & (x,y) &= (0,r) \\ t &= \frac{\pi}{4}, & (x,y) &= (r,0) \\ t &= \frac{\pi}{2}, & (x,y) &= (0,-r) \\ t &= \frac{3\pi}{4}, & (x,y) &= (-r,0) \\ t &= \pi, & (x,y) &= (0,r) \end{aligned}$$

## 5.7. PARAMETRIC EQUATIONS

(e) Since

$$x = r \sin 3t \quad \text{and} \quad y = r \cos 3t, \quad \text{for } 0 \leq t \leq 2\pi,$$

if

$$\begin{aligned} t &= 0, & (x,y) &= (0,r) \\ t &= \frac{\pi}{6}, & (x,y) &= (r,0) \\ t &= \frac{\pi}{3}, & (x,y) &= (0,-r) \\ t &= \frac{\pi}{2}, & (x,y) &= (-r,0) \\ t &= \frac{2\pi}{3}, & (x,y) &= (0,r) \end{aligned}$$

■

### 5.7.1 Using Graphing Devices

**Example 216** *(Exercise Set 5.7, Exercise 29) A cycloid can be described as the path traced out by a point on a circle as the circle rolls along a line. If a is the radius of the circle, the parametric equations of the cycloid are given by*

$$x = a(t - \sin t), \quad y = a(1 - \cos t).$$

*Use a graphing device to sketch the cycloid for different values of a. What effect does the parameter a have on the curve? For what values of t does the curve touch the x-axis?*

Solution: The figure shows the curve for $a = 1, 2,$ and $3$.

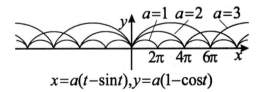

$x=a(t-\sin t), y=a(1-\cos t)$

The parameter $a$ determines the maximum height of the bumps in the curve, which have a height of $2a$. The parameter $a$ also determines the period of the curve, which is $2\pi a$, and equals the circumference of the rolling circle.

To find the locations where the curve touches the $x$-axis, solve for those values of $t$ that make the $y$-coordinate 0. So

$$y = a(1 - \cos t) = 0$$
$$1 - \cos t = 0$$
$$\cos t = 1$$
$$t = 0, \pm 2\pi, \pm 4\pi, \ldots$$
$$t = \pm 2k\pi, k = 0, 1, 2, \ldots$$
$$x = a(t - \sin t) = at = \pm 2ak\pi, \quad \text{for } k = 0, 1, 2, \ldots$$

■

## 5.8 Rotation of Axes

The *general equation of a conic section* is

$$Ax^2 + Bxy + Cy^2 + Dx + Ey + F = 0.$$

If $B = 0$, then the conic section

$$Ax^2 + Cy^2 + Dx + Ey + F = 0$$

is a parabola when $AC = 0$, is an ellipse when $AC > 0$ and is a hyperbola when $AC < 0$. When $B \neq 0$, the curve is still a conic but with a rotated axis. For every such equation, there is an angle $\theta$, with $0 < \theta < \frac{\pi}{2}$, so that if the $xy$-coordinate system is rotated through this angle to form a new $\hat{x}\hat{y}$-coordinate system, in the new coordinates the curve has an equation without an $\hat{x}\hat{y}$ term, allowing the curve to be analyzed as before.

If a rectangular $xy$-coordinate system is rotated through an angle $\theta$ to form an $\hat{x}\hat{y}$-coordinate system, then the coordinates of a point $(x, y)$ in the $xy$-plane and the coordinates of a point $(\hat{x}, \hat{y})$ in the $\hat{x}\hat{y}$-plane are related by the formulas

$$x = \hat{x}\cos\theta - \hat{y}\sin\theta \quad \text{and} \quad y = \hat{x}\sin\theta + \hat{y}\cos\theta$$
$$\hat{x} = x\cos\theta + y\sin\theta \quad \text{and} \quad \hat{y} = -x\sin\theta + y\cos\theta.$$

## 5.8. ROTATION OF AXES

To eliminate the $xy$-term in the general *quadratic equation*

$$Ax^2 + Bxy + Cy^2 + Dx + Ey + F = 0, \text{ where } B \neq 0$$

rotate the coordinate axes through an angle $\theta$ that satisfies

$$\cot 2\theta = \frac{A-C}{B}.$$

If the new equation is denoted

$$\hat{A}x^2 + \hat{C}y^2 + \hat{D}x + \hat{E}y + \hat{F} = 0$$

then the conic $Ax^2 + Bxy + Cy^2 + Dx + Ey + F = 0$ is:
1. An ellipse if $B^2 - 4AC = -4\hat{A}\hat{C} < 0$;
2. A hyperbola if $B^2 - 4AC = -4\hat{A}\hat{C} > 0$;
3. A parabola if $B^2 - 4AC = -4\hat{A}\hat{C} = 0$.

**Example 217** *(Exercise Set 5.8, Exercise 5) Determine whether the conic*

$$x^2 - xy + y^2 = 2$$

*is an ellipse, hyperbola, or parabola, and perform a rotation, and if necessary a translation, and sketch the graph.*

Solution: The equation is a general quadratic with

$$A = 1, \quad B = -1, \quad C = 1, \quad D = E = 0, \quad F = -2$$

so

$$B^2 - 4AC = 1 - 4(1)(1) = -3 < 0$$

and the conic is an ellipse.

To eliminate the $xy$ term rotate the coordinate axes through an angle $\theta$ satisfying

$$\cot 2\theta = \frac{A-C}{B}$$
$$= \frac{1-1}{-1} = 0$$

so

$$\cot 2\theta = 0$$
$$\frac{\cos 2\theta}{\sin 2\theta} = 0$$
$$\cos 2\theta = 0$$
$$2\theta = \frac{\pi}{2}, \quad \theta = \frac{\pi}{4}.$$

To rewrite the equation in the new $\hat{x}\hat{y}$-plane use the formulas

$$x = \hat{x}\cos\theta - \hat{y}\sin\theta \quad \text{and} \quad y = \hat{x}\sin\theta + \hat{y}\cos\theta$$

to determine $x$ and $y$ in terms of $\hat{x}$ and $\hat{y}$. So

$$x = \hat{x}\cos\frac{\pi}{4} - \hat{y}\sin\frac{\pi}{4} \quad \text{and} \quad y = \hat{x}\sin\frac{\pi}{4} + \hat{y}\cos\frac{\pi}{4}$$
$$x = \frac{\sqrt{2}}{2}\hat{x} - \frac{\sqrt{2}}{2}\hat{y} \quad \text{and} \quad y = \frac{\sqrt{2}}{2}\hat{x} + \frac{\sqrt{2}}{2}\hat{y}$$
$$x = \frac{\sqrt{2}}{2}(\hat{x} - \hat{y}) \quad \text{and} \quad y = \frac{\sqrt{2}}{2}(\hat{x} + \hat{y}).$$

Now substitute the expressions for $x$ and $y$ into the original equation. This gives

$$\left[\frac{\sqrt{2}}{2}(\hat{x} - \hat{y})\right]^2 - \left[\frac{\sqrt{2}}{2}(\hat{x} - \hat{y})\right]\left[\frac{\sqrt{2}}{2}(\hat{x} + \hat{y})\right] + \left[\frac{\sqrt{2}}{2}(\hat{x} + \hat{y})\right]^2 = 2$$
$$\frac{1}{2}(\hat{x} - \hat{y})^2 - \frac{1}{2}(\hat{x} - \hat{y})(\hat{x} + \hat{y}) + \frac{1}{2}(\hat{x} + \hat{y})^2 = 2$$
$$\frac{1}{2}\hat{x}^2 + \frac{3}{2}\hat{y}^2 = 2$$
$$\hat{x}^2 + 3\hat{y}^2 = 4.$$

The new equation is the graph of an ellipse in standard form in the $\hat{x}\hat{y}$-plane.

## 5.8. ROTATION OF AXES

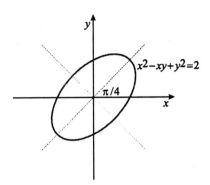

**Example 218** *(Exercise Set 5.8, Exercise 11)* Determine whether the conic

$$6x^2 + 4\sqrt{3}xy + 2y^2 - 9x + 9\sqrt{3}y - 63 = 0$$

*is an ellipse, hyperbola, or parabola, and perform a rotation, and if necessary a translation, and sketch the graph.*

Solution:
Classification: Since

$$A = 6, B = 4\sqrt{3}, \text{ and } C = 2$$

we have

$$B^2 - 4AC = 48 - 4(6)(2) = 0$$

so the conic is a parabola.
Eliminate the $xy$-term: Rotate through an angle $\theta$ satisfying

$$\begin{aligned} \cot 2\theta &= \frac{A-C}{B} \\ &= \frac{6-2}{4\sqrt{3}} = \frac{4}{4\sqrt{3}} = \frac{1}{\sqrt{3}} \\ &= \frac{\sqrt{3}}{3}. \end{aligned}$$

Since

$$0 < \theta < \frac{\pi}{2} \Rightarrow 0 < 2\theta < \pi,$$

and since $\sin 2\theta > 0$ whenever $0 < 2\theta < \pi$, in order for the cotangent to be positive, we must have $\cos 2\theta > 0$. The figure shows a right triangle with one angle $2\theta$ and opposite and adjacent sides 3 and $\sqrt{3}$, respectively. So $\cot 2\theta = \frac{\sqrt{3}}{3}$.

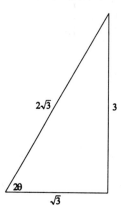

The hypotenuse of the right triangle is determined using the Pythagorean Theorem and is

$$\sqrt{\left(\sqrt{3}\right)^2 + 3^2} = \sqrt{12} = 2\sqrt{3}.$$

Then from the triangle we have

$$\cos 2\theta = \frac{1}{2}.$$

From our knowledge of the values of the trigonometric functions, the angle is found for

$$2\theta = 60° = \frac{\pi}{3}, \quad \text{so } \theta = 30° = \frac{\pi}{6}.$$

Substitute:

$$\cos\theta = \cos\frac{\pi}{6} = \frac{\sqrt{3}}{2}, \sin\theta = \sin\frac{\pi}{6} = \frac{1}{2}$$

then

$$x = \hat{x}\cos\theta - \hat{y}\sin\theta = \frac{\sqrt{3}}{2}\hat{x} - \frac{1}{2}\hat{y} = \frac{1}{2}(\sqrt{3}\hat{x} - \hat{y})$$

$$y = \hat{x}\sin\theta + \hat{y}\cos\theta = \frac{1}{2}\hat{x} + \frac{\sqrt{3}}{2}\hat{y} = \frac{1}{2}(\hat{x} + \sqrt{3}\hat{y})$$

## 5.8. ROTATION OF AXES

So the new $\hat{x}\hat{y}$-equation is

$$6\left[\frac{1}{2}(\sqrt{3}\hat{x}-\hat{y})\right]^2 + 4\sqrt{3}\left[\frac{1}{2}(\sqrt{3}\hat{x}-\hat{y})\right]\left[\frac{1}{2}\hat{x}+\sqrt{3}\hat{y})\right] + 2\left[\frac{1}{2}(\hat{x}+\sqrt{3}\hat{y})\right]^2$$
$$-9\left[\frac{1}{2}(\sqrt{3}\hat{x}-\hat{y})\right] + 9\sqrt{3}\left[\frac{1}{2}(\hat{x}+\sqrt{3}\hat{y})\right] - 63 = 0$$
$$\frac{3}{2}(\sqrt{3}\hat{x}-\hat{y})^2 + \sqrt{3}(\sqrt{3}\hat{x}-\hat{y})(\hat{x}+\sqrt{3}\hat{y}) + \frac{1}{2}(\hat{x}+\sqrt{3}\hat{y})^2 - \frac{9}{2}(\sqrt{3}\hat{x}-\hat{y})$$
$$+\frac{9\sqrt{3}}{2}(\hat{x}+\sqrt{3}\hat{y}) - 63 = 0.$$

The equation simplifies to

$$8\hat{x}^2 + 18\hat{y} - 63 = 0$$

or to

$$\hat{y} = -\frac{4}{9}\hat{x}^2 + \frac{7}{2},$$

which is the equation of a parabola in the $\hat{x}\hat{y}$-plane. Notice that the parabola can be obtained from the parabola $\hat{y} = \hat{x}^2$, by first reflection it over the $\hat{x}$-axis, scaling the resulting graph by a factor of $\frac{4}{9}$, and then shifting the graph downward $\frac{7}{2}$ units.

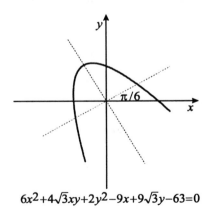

$6x^2+4\sqrt{3}xy+2y^2-9x+9\sqrt{3}y-63=0$

∎

**Example 219** *Determine whether the conic*

$$108x^2 - 312xy + 17y^2 + 240x + 320y + 500 = 0$$

is an ellipse, hyperbola, or parabola, and perform a rotation, and if necessary a translation, and sketch the graph.

Solution:
<u>Classification:</u> $A = 108, B = -312, C = 17$

$$B^2 - 4AC = (-312)^2 - 4(108)(17) = 90000 > 0$$

so the conic is a hyperbola.
<u>Eliminate the $xy$-term:</u> Rotate through an angle $\theta$ satisfying

$$\cot 2\theta = \frac{A - C}{B} = \frac{108 - 17}{-312} = -\frac{7}{24}.$$

The figure shows a right triangle with one angle $\varphi$ whose opposite and adjacent sides are 24 and 7. So

$$\cot \varphi = \frac{7}{24}.$$

By the Pythagorean Theorem the hypotenuse is 25, so

$$\cos \varphi = \frac{7}{25} \Rightarrow \cos 2\theta = -\frac{7}{25}.$$

In this case the angle $2\theta$ can not be exactly determined from the value of the cosine. However, the half angle formulas allow us to compute the $\sin \theta$ and $\cos \theta$ from the $\cos 2\theta$, which is all that is needed. So

## 5.8. ROTATION OF AXES

$$\cos\theta = \sqrt{\frac{1+\cos 2\theta}{2}} = \sqrt{\frac{1-\frac{7}{25}}{2}} = \frac{3}{5}$$

$$\sin\theta = \sqrt{\frac{1-\cos 2\theta}{2}} = \sqrt{\frac{1+\frac{7}{25}}{2}} = \frac{4}{5}.$$

Substitute: $\cos\theta = \frac{3}{5}, \sin\theta = \frac{4}{5}$

$$x = \hat{x}\cos\theta - \hat{y}\sin\theta = \frac{3}{5}\hat{x} - \frac{4}{5}\hat{y} = \frac{1}{5}(3\hat{x} - 4\hat{y})$$

$$y = \hat{x}\sin\theta + \hat{y}\cos\theta = \frac{4}{5}\hat{x} + \frac{3}{5}\hat{y} = \frac{1}{5}(4\hat{x} + 3\hat{y})$$

The $\hat{x}\hat{y}$ equation is therefore

$$\tfrac{108}{25}(3\hat{x}-4\hat{y})^2 - \frac{312}{25}(3\hat{x}-4\hat{y})(4\hat{x}+3\hat{y}) + \frac{17}{25}(4\hat{x}+3\hat{y})^2$$
$$+\frac{240}{5}(3\hat{x}-4\hat{y}) + \frac{320}{5}(4\hat{x}+3\hat{y}) + 500 = 0,$$

which simplifies to

$$-100\hat{x}^2 + 225\hat{y}^2 + 400\hat{x} + 500 = 0.$$

To place the equation in a standard form by completing the square on the $\hat{x}$ terms. This gives

$$\begin{aligned}
-100\hat{x}^2 + 225\hat{y}^2 + 400\hat{x} + 500 &= 0 \\
100(\hat{x}^2 - 4\hat{x}) - 225\hat{y}^2 &= 500 \\
100(\hat{x}^2 - 4\hat{x} + 4) - 225\hat{y}^2 &= 500 + 400 \\
100(\hat{x} - 2)^2 - 225\hat{y}^2 &= 900 \\
\frac{(\hat{x}-2)^2}{9} - \frac{\hat{y}^2}{4} &= 1.
\end{aligned}$$

Notice that the center of the hyperbola is shifted to $(2, 0)$ in the $\hat{x}\hat{y}$-plane.

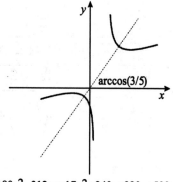

$180x^2 - 312xy + 17y^2 + 240x + 320y + 500 = 0$

■

# APPENDIX

Two copies of a placement examination of the type that is used at Youngstown State University are included in this Appendix. They can also be down loaded from the WWW site

http://www.cis.ysu.edu/a_s/mathematics/JDFaires/precalculus

We suggest that you work one of these examinations before you take your Precalculus course and the other after you complete the course. If you score 16 or higher on the 40 question examination we would expect that you are sufficiently prepared to take a Precalculus course based on this book. A score of 28 would indicate to us that you are well-prepared for a University Calculus sequence.

# PRECALCULUS-CALCULUS READINESS EXAMINATION

## Test # 1

1. When $a = -2$, $b = 3$, and $c = 4$, the value of the expression $\dfrac{b^2 - c}{a^2 - b^2}$ is

    (a) 1  (b) $-1$  (c) $-\dfrac{5}{13}$  (d) $\dfrac{5}{13}$
    (e) None of the above.

2. The expression $(-3b)\left(-2a^2 b\right)^2$ simplifies to

    (a) $-12a^4 b^3$  (b) $36a^4 b^4$  (c) $12a^4 b^3$  (d) $6a^2 b^2$
    (e) None of the above.

3. The expression $6x^2 - 7[-2x^3 - 6(3x^3 - 1)]$ simplifies to

    (a) $112x^3 - 6x^2 - 42$  (b) $140x^3 + 6x^2 - 42$  (c) $-140x^3 + 6x^2 + 42$  (d) $-112x^3 + 6x^2 - 42$
    (e) None of the above.

4. If $3(x+2) + x = 4\left(1 - \dfrac{1}{2}x\right)$, then $x$ is

    (a) $-\dfrac{1}{2}$  (b) $-\dfrac{1}{3}$  (c) $0$  (d) $-1$
    (e) None of the above.

5. The expression $\dfrac{\sqrt{x}\,\sqrt[3]{x}}{\sqrt[4]{x}}$ simplifies to

    (a) $x$  (b) $x^{29/6}$  (c) $x^{-1}$  (d) $x^{7/12}$
    (e) None of the above.

6. The quotient $\dfrac{\dfrac{1}{x+h}-\dfrac{1}{x}}{h}$ simplifies to

   (a) $\dfrac{1}{h^2}$  (b) $\dfrac{-h^2}{(x+h)x}$  (c) $-\dfrac{1}{x(x+h)}$  (d) $\dfrac{1}{x(x+h)}$
   (e) None of the above.

7. If $-6x+1<2$, then

   (a) $x>-\dfrac{1}{6}$  (b) $x>-\dfrac{1}{2}$  (c) $x<-\dfrac{1}{6}$  (d) $x<-\dfrac{2}{5}$
   (e) None of the above.

8. If $\dfrac{7}{5x-6}=\dfrac{1}{x-1}$, then $x$ is

   (a) $\dfrac{1}{12}$  (b) $\dfrac{1}{2}$  (c) $-\dfrac{5}{2}$  (d) $-\dfrac{13}{2}$
   (e) None of the above.

9. The quadratic expression $x^2-19x+84$ factors as

   (a) $(x-12)(x-7)$  (b) $(x+12)(x+7)$  (c) $(x-21)(x-4)$  (d) $(x-28)(x-3)$
   (e) None of the above.

10. The expression $\dfrac{4x-5}{x^2-3x+2}-\dfrac{3}{x-2}$ simplifies to

    (a) $\dfrac{3}{x-1}$  (b) $\dfrac{x-8}{(x-2)(x+1)}$  (c) $\dfrac{1}{x-1}$  (d) $\dfrac{x-8}{(x-2)(x-1)}$
    (e) None of the above.

11. The polynomial $24x-14x^2-3x^3$ can be factored as

    (a) $-x(x+6)(3x-4)$  (b) $-3x(x+4)(x-2)$  (c) $-x(3x-8)(x+3)$  (d) $-x(x-6)(3x+4)$
    (e) None of the above.

12. The expression $\dfrac{x^2 - 9}{x + 3} \cdot \dfrac{2x - 6}{x^2 - 6x + 9}$ simplifies to

(a) 1    (b) $\dfrac{1}{3x}$    (c) $\dfrac{2}{x + 3}$    (d) 2

(e) None of the above.


13. The quadratic equation $x^2 - x - 7 = 0$ is satisfied when $x$ is

(a) $1 \pm 2\sqrt{2}$    (b) $\dfrac{1}{2}(1 \pm 2\sqrt{2})$    (c) $\dfrac{1}{2}(1 \pm \sqrt{29})$    (d) $\dfrac{1}{2}(-1 \pm \sqrt{29})$

(e) None of the above.


14. If $|x - 3| \leq 4$, then

(a) $1 \leq x \leq 7$    (b) $-7 \leq x \leq -1$    (c) $-1 \leq x \leq 7$    (d) $-7 \leq x \leq 1$

(e) None of the above.


15. One factor of $25y^{16} - 4x^2$ is

(a) $5y^8 + 4x$    (b) $5y^{12} - 2x$    (c) $5y^4 + 2x$    (d) $5y^8 - 2x$

(e) None of the above.


16. The inequality $(2x - 3)(x - 4) < 0$ is satisfied when

(a) $\dfrac{3}{2} < x < 4$    (b) $x > 4$ only    (c) $x < \dfrac{3}{2}$ only    (d) $x < \dfrac{3}{2}$ or $x > 4$

(e) None of the above.


17. The expression $\left(\dfrac{x^8 y^{-4}}{z^{-8}}\right)^{-\frac{3}{4}}$ is equivalent to

(a) $x^6 y^3 z^{-6}$    (b) $x^{-6} y^{-3} z^6$    (c) $x^{-6} y^3 z^6$    (d) $x^{-6} y^3 z^{-6}$

(e) None of the above.

18. The slope of the line with equation $4x + 3 = y - 2$ is

    (a) $\frac{4}{5}$　　　(b) $\frac{4}{3}$　　　(c) $-\frac{4}{3}$　　　(d) 4
    
    (e) None of the above.

19. A line with slope 2 and y-intercept 3 has the equation

    (a) $y = 3x + 2$　　(b) $x = 2y + 3$　　(c) $y = 2x + 3$　　(d) $x = 3y + 2$
    
    (e) None of the above.

20. If $(x,y)$ satisfies both of the equations $2x + 3y = 7$ and $2x - 3y = 4$, then $x$ is

    (a) $\frac{3}{4}$　　　(b) 5　　　(c) $\frac{5}{4}$　　　(d) $\frac{11}{4}$
    
    (e) None of the above.

21. If $f(x) = 2 - 3x$, then $f(x-1) + f(x) - 1$ is

    (a) $2 - 6x$　　(b) $-6x$　　(c) $6 - 6x$　　(d) $8 - 6x$
    
    (e) None of the above.

22. The largest set of real numbers in the domain of the function $f(x) = \frac{1}{\sqrt{3-x}}$ is

    (a) $x < 3$　　(b) $x \leq 3$　　(c) $x < -3$　　(d) $x \leq -3$
    
    (e) None of the above.

23. If $f(x) = 3x - 5$ and $g(x) = 2x^2$, then the composition $(f \circ g)(x) \equiv f(g(x))$ is

    (a) $6x^2 - 5$　　(b) $2x^2 - 5$　　(c) $18x^2 - 60x + 50$　　(d) $6x^3 - 10x^2$
    
    (e) None of the above.

273

24. The equation $\dfrac{x^2 - 3x + 2}{x - 1} + 2x = 4$ is satisfied when $x$ is

(a) 1 or 2  (b) $\dfrac{2}{3}$ only  (c) 2 only  (d) $-1$ or $\dfrac{2}{3}$

(e) None of the above.


25. The expression $\dfrac{(1-x)(x+2) + (x+2)^2}{(1-x)(x+2)}$ simplifies to

(a) $x^2 + 4x + 4$  (b) $x^2 + 4x + 5$  (c) $x + 3$  (d) $\dfrac{3}{1-x}$

(e) None of the above.


26. The distance between the points $P(3, 8)$ and $Q(-5, 2)$ is

(a) 14  (b) 18  (c) $2\sqrt{10}$  (d) 10

(e) None of the above.


27. For positive real numbers $m$, $n$, and $r$, which of the following are true?

   I. $\log(mn) = \log m + \log n$
   
   II. $\log\left(\dfrac{m}{n}\right) = \dfrac{\log m}{\log n}$
   
   III. $\log(m^r) = r \log m$

(a) I, II, and III  (b) I and II  (c) I and III  (d) II and III

(e) None of the above.


28. Suppose that $3^8$ is approximately $7{,}000$, which of the following best approximates $3^{16}$?

(a) $(7{,}000)^8$  (b) $490{,}000$  (c) $14{,}000$  (d) $49{,}000{,}000$

(e) None of the above.

29. A rectangle has a length that is 6 meters more than its width. What is the width of the rectangle if the perimeter of the rectangle is 156 meters?

   (a) 36 meters    (b) 42 meters    (c) 75 meters    (d) $-3+\sqrt{165}$ meters

   (e) None of the above.

30. The volume of a sphere is proportional to the cube of its radius. Suppose the sphere $S$ has a radius that is twice the radius of the sphere $C$ and that the volume of $C$ is $V$. Then the volume of $S$ is

   (a) $2V$    (b) $3V$    (c) $6V$    (d) $8V$

   (e) None of the above.

31. The graph of $y = \cos\frac{1}{3}x$, for $x$ in the interval $[0, 3\pi]$, crosses the $x$-axis at

   (a) $0, \frac{3\pi}{2}$, and $3\pi$    (b) $0$ and $3\pi$    (c) $\frac{3\pi}{2}$ only    (d) $0$ only

   (e) None of the above.

32. When the expression $\dfrac{1}{\tan t + \cot t}$ is defined, it is equivalent to

   (a) $1$    (b) $\dfrac{1}{\sin t \cos t}$    (c) $\sin t \cos t$    (d) $\dfrac{\sin t \cos t}{\sin t + \cos t}$

   (e) None of the above.

33. The solutions of the equation $2\cos x - \sqrt{3} = 0$ that lie in the interval $[0, 2\pi]$ are

   (a) $\frac{\pi}{6}$ and $\frac{11\pi}{6}$    (b) $\frac{\pi}{6}$ and $\frac{5\pi}{6}$    (c) $\frac{\pi}{3}$ and $\frac{5\pi}{3}$    (d) $\frac{\pi}{3}$ and $\frac{2\pi}{3}$

   (e) None of the above.

34. The value of $\sin(t - \pi)$ is the same as the value of

   (a) $-\sin t$    (b) $\sin t$    (c) $-\cos t$    (d) $\cos t$

   (e) None of the above.

35. An open rectangular box has height $h$ and a square base with a perimeter $4x$. The surface area of the box is

(a) $x^2 h$  (b) $4x + h$  (c) $2x^2 + 4xh$  (d) $x^2 + 4xh$

(e) None of the above.

36. In an isosceles right triangle two sides have length 5. The length of the hypotenuse is

(a) $\sqrt{10}$  (b) $\dfrac{5\sqrt{3}}{2}$  (c) $5\sqrt{2}$  (d) $5\sqrt{3}$

(e) None of the above.

37. An angle of $\dfrac{7\pi}{3}$ radians is the same as an angle of

(a) 180°  (b) 420°  (c) 240°  (d) 30°

(e) None of the above.

38. A rectangle $R$ has width $x$ and length $y$. A new rectangle $S$ is formed from $R$ by multiplying all the sides of $R$ by 6. How much more area does $S$ have than $R$?

(a) $5xy$  (b) $6xy$  (c) $35xy$  (d) $36xy$

(e) None of the above.

39. The equation $x^2 + y^2 + 6x - 2y - 15 = 0$ describes a circle with

(a) center $(-3, 1)$ and radius 5
(b) center $(3, -1)$ and radius 5
(c) center $(-3, 1)$ and radius 25
(d) center $(3, -1)$ and radius 25

(e) None of the above.

40. The graph of the equation $y = \dfrac{2x}{x^2 - 2}$ has a vertical asymptote whose equation is

(a) $y = 2$  (b) $x = -\sqrt{2}$  (c) $y = \sqrt{2}$  (d) $x = 2$

(e) None of the above.

# PRECALCULUS-CALCULUS READINESS EXAMINATION

## Test # 2

1. When $a = 2$, $b = -3$, and $c = 5$, the value of the expression $\dfrac{a^2 - b}{c^2 - b^2}$ is

   (a) $\dfrac{1}{3}$   (b) $\dfrac{7}{34}$   (c) $\dfrac{7}{16}$   (d) $\dfrac{1}{16}$

   (e) None of the above.

2. The expression $(-2a)\left(-4a^2b^3\right)^2$ simplifies to

   (a) $16a^5b^6$   (b) $-32a^5b^5$   (c) $64a^6b^6$   (d) $-32a^5b^6$

   (e) None of the above.

3. The expression $3x^2 + 5[4x^2 - 6(3x + 1)]$ simplifies to

   (a) $23x^2 - 90x + 30$   (b) $23x^2 - 90x + 5$   (c) $23x^2 - 90x - 30$   (d) $23x^2 - 18x + 1$

   (e) None of the above.

4. If $2(x + 3) + x = 6\left(1 - \dfrac{1}{3}x\right)$, then $x$ is

   (a) $\dfrac{3}{5}$   (b) $-\dfrac{3}{8}$   (c) $0$   (d) $\dfrac{1}{5}$

   (e) None of the above.

5. The expression $\dfrac{\sqrt[4]{x}\,\sqrt[3]{x}}{x}$ simplifies to

   (a) $x^{-5/12}$   (b) $x^{-8}$   (c) $x^{5/12}$   (d) $x^{19/12}$

   (e) None of the above.

277

6. The quotient $\dfrac{(x+h)^2 - x^2}{h}$ simplifies to

   (a) $h$  (b) $2x + h$  (c) $x + h$  (d) $2xh + h$
   
   (e) None of the above.

7. If $-5x + 1 < 5$, then

   (a) $x > -\dfrac{4}{5}$  (b) $x < -\dfrac{4}{5}$  (c) $x > -\dfrac{6}{5}$  (d) $x < -\dfrac{5}{4}$
   
   (e) None of the above.

8. If $\dfrac{5}{3x-4} = \dfrac{1}{x-2}$, then $x$ is

   (a) $-1$  (b) $3$  (c) $\dfrac{3}{4}$  (d) $1$
   
   (e) None of the above.

9. The quadratic expression $x^2 - 7x + 10$ factors as

   (a) $(x+2)(x+5)$  (b) $(x-2)(x-5)$  (c) $(x+1)(x+10)$  (d) $(x-1)(x-10)$
   
   (e) None of the above.

10. The expression $\dfrac{3x-4}{x^2-5x+6} - \dfrac{1}{x-3}$ simplifies to

    (a) $\dfrac{2(x-1)}{(x-2)(x-3)}$  (b) $\dfrac{2}{(x-2)}$  (c) $\dfrac{2(x-5)}{(x+2)(x-3)}$  (d) $\dfrac{2(2x-5)}{(x-2)(x-3)}$
    
    (e) None of the above.

11. The polynomial $x^3 + 7x^2 + 6x$ can be factored as

    (a) $x(x-1)(x-6)$  (b) $x(x-2)(x-3)$  (c) $x(x+2)(x+3)$  (d) $x(x+1)(x+6)$
    
    (e) None of the above.

12. The expression $\dfrac{x^2-16}{x+4} \cdot \dfrac{3x-12}{x^2-8x+16}$ simplifies to

(a) 1  (b) 0  (c) 3  (d) $-\dfrac{3}{8x}$

(e) None of the above.

13. The quadratic equation $x^2 - x - 5 = 0$ is satisfied when $x$ is

(a) $1 \pm \sqrt{6}$  (b) $\dfrac{1}{2}(1 \pm \sqrt{21})$  (c) $-\dfrac{1}{2}(1 \pm \sqrt{21})$  (d) $\dfrac{1}{2}(1 \pm \sqrt{6})$

(e) None of the above.

14. If $|x-2| \leq 5$, then

(a) $3 \leq x \leq 7$  (b) $-7 \leq x \leq -3$  (c) $-3 \leq x \leq 7$  (d) $-7 \leq x \leq 3$

(e) None of the above.

15. One factor of $9y^6 - 25x^2$ is

(a) $3y^3 + 5x$  (b) $9y^4 - 5x$  (c) $y^3 - 5x$  (d) $3y^4 + 25x$

(e) None of the above.

16. The inequality $\dfrac{x-2}{x+5} > 0$ is satisfied when

(a) $x < -5$ only  (b) $x < -5$ or $x > 2$  (c) $x < 2$ or $x > 5$  (d) $x > 2$ only

(e) None of the above.

17. The expression $\left(\dfrac{x^{12}y^{-3}}{z^{-3}}\right)^{-\frac{4}{3}}$ is equivalent to

(a) $x^{-16}y^{-4}z^4$  (b) $x^{-16}y^4z^4$  (c) $x^{16}y^4z^{-4}$  (d) $x^{-16}y^4z^{-4}$

(e) None of the above.

18. The slope of the line with equation $3x + 4 = y - 5$ is

(a) $\frac{3}{5}$  (b) 3  (c) $\frac{3}{4}$  (d) $-\frac{3}{5}$

(e) None of the above.

19. A line with slope 3 and y-intercept 5 has the equation

(a) $y = 5x + 3$  (b) $x = 5y + 3$  (c) $y = 3x + 5$  (d) $x = 3y + 5$

(e) None of the above.

20. If $(x,y)$ satisfies both of the equations $3x + 4y = 9$ and $3x - 4y = 6$, then $x$ is

(a) $\frac{5}{2}$  (b) $\frac{1}{2}$  (c) $\frac{3}{2}$  (d) $\frac{9}{2}$

(e) None of the above.

21. If $f(x) = 2 - 4x$, then $f(x-2) + f(x) - 2$ is

(a) $14 - 8x$  (b) $10 - 8x$  (c) $-8x$  (d) $10 + 8x$

(e) None of the above.

22. The largest set of real numbers in the domain of the function $f(x) = \dfrac{1}{\sqrt{27 - x^3}}$ is

(a) $x < -3$  (b) $x \leq -3$  (c) $x < 3$  (d) $x \leq 3$

(e) None of the above.

23. If $f(x) = 2x - 3$ and $g(x) = 8x^2$, then the composition $(f \circ g)(x) \equiv f(g(x))$ is

(a) $8x^2 - 3$  (b) $32x^2 - 96x - 72$  (c) $16x^2 - 3$  (d) $16x^3 - 24x^2$

(e) None of the above.

24. The equation $\dfrac{x^2 - 5x + 6}{x - 2} + x = 3$ is satisfied when $x$ is

(a) 3 only  (b) 2 or 3  (c) 0 only  (d) 2 or $\dfrac{3}{2}$

(e) None of the above.

25. The expression $\dfrac{(2-x)(x+3) + (x+3)(x-4)}{(x+3)(2-x)}$ simplifies to

(a) $x^2 - x - 11$  (b) $x^2 - x - 12$  (c) $\dfrac{2}{x-2}$  (d) $x - 3$

(e) None of the above.

26. The distance between the points $P(4,7)$ and $Q(-1,-5)$ is

(a) 5  (b) 17  (c) 13  (d) $\sqrt{13}$

(e) None of the above.

27. For positive real numbers $m$, $n$, and $r$, which of the following are true?

  I.   $\log(mn) = (\log m)(\log n)$
  II.  $\log\left(\dfrac{m}{n}\right) = \log m - \log n$
  III. $\log(m^r) = r \log m$

(a) I, II, and III  (b) I and II  (c) I and III  (d) II and III

(e) None of the above.

28. Suppose that $2^{11}$ is approximately $2{,}000$, which of the following best approximates $2^{22}$?

(a) $(4{,}000)^{11}$  (b) $40{,}000$  (c) $4{,}000{,}000$  (d) $(2{,}000)^{11}$

(e) None of the above.

29. A rectangle has a length that is 2 meters more than its width. What is the width of the rectangle if the perimeter of the rectangle is 52 meters?

   (a) $-1+\sqrt{53}$ meters  (b) 25 meters  (c) 14 meters  (d) 12 meters
   (e) None of the above.

30. The volume of a sphere is proportional to the cube of its radius, and the surface area of a sphere is proportional to the square of the radius. Suppose the sphere $S$ has a surface area that is 4 times the surface area of $C$. If the volume of $C$ is 27, then the volume of $S$ is

   (a) 108  (b) 162  (c) 216  (d) 243
   (e) None of the above.

31. The graph of $y = \sin\frac{1}{3}x$, for $x$ in the interval $[0, 3\pi]$, crosses the $x$-axis at

   (a) $0, \frac{3\pi}{2}$, and $3\pi$  (b) 0 and $3\pi$  (c) $\frac{3\pi}{2}$ only  (d) 0 only
   (e) None of the above.

32. When the expression $\sin t(\tan t + \cot t)$ is defined, it is equivalent to

   (a) 1  (b) $\frac{1}{\sin t}$  (c) $\frac{1}{\cos t}$  (d) $(\sin t)^2 \cos t$
   (e) None of the above.

33. The solutions of the equation $2\cos x - \sqrt{2} = 0$ that lie in the interval $[0, 2\pi]$ are

   (a) $\frac{\pi}{4}$ and $\frac{3\pi}{4}$  (b) $\frac{\pi}{4}$ and $\frac{5\pi}{4}$  (c) $\frac{\pi}{4}$ and $\frac{7\pi}{4}$  (d) $\frac{3\pi}{4}$ and $\frac{7\pi}{4}$
   (e) None of the above.

34. The value of $\tan(t - \pi)$ is the same as the value of

   (a) $\tan t$  (b) $-\tan t$  (c) $\cot t$  (d) $-\cot t$
   (e) None of the above.

35. An open rectangular box has height 3 and a square base. The volume of the box is 48. The surface area of the box is

   (a) 32      (b) 64      (c) 80      (d) $24\sqrt[3]{6} + 8\sqrt[3]{36}$

   (e) None of the above.

36. In an isosceles right triangle two sides have length 7. The length of the hypotenuse is

   (a) $7\sqrt{2}$      (b) $\dfrac{7\sqrt{3}}{2}$      (c) $\sqrt{14}$      (d) $7\sqrt{3}$

   (e) None of the above.

37. An angle of $\dfrac{8\pi}{3}$ radians is the same as an angle of

   (a) 480°      (b) 420°      (c) 240°      (d) 60°

   (e) None of the above.

38. A square $R$ has perimeter 8. A new square $S$ is formed from $R$ by multiplying all the sides of $R$ by 3. How much more area does $S$ have than $R$?

   (a) 8      (b) 32      (c) 36      (d) 512

   (e) None of the above.

39. The equation $x^2 + y^2 - 6x - 2y - 15 = 0$ describes a circle with

   (a) center $(-3, -1)$ and radius 5      (b) center $(3, 1)$ and radius 5
   (c) center $(-3, -1)$ and radius 25     (d) center $(3, 1)$ and radius 25

   (e) None of the above.

40. The graph of the equation $y = \dfrac{3x}{x^2 - 9}$ has a vertical asymptote whose equation is

   (a) $x = 3$      (b) $y = 3$      (c) $x = 9$      (d) $y = -3$

   (e) None of the above.

# PRECALCULUS-CALCULUS READINESS TEST ANSWERS

| Problem | Test 1 | Test 2 | Problem | Test 1 | Test 2 |
|---|---|---|---|---|---|
| 1 | b | c | 21 | c | b |
| 2 | a | d | 22 | a | c |
| 3 | b | c | 23 | a | c |
| 4 | b | c | 24 | c | a |
| 5 | d | a | 25 | d | c |
| 6 | c | b | 26 | d | c |
| 7 | a | a | 27 | c | d |
| 8 | b | b | 28 | d | c |
| 9 | a | b | 29 | a | d |
| 10 | c | a | 30 | d | c |
| 11 | a | d | 31 | c | b |
| 12 | d | c | 32 | c | c |
| 13 | c | b | 33 | a | c |
| 14 | c | c | 34 | a | a |
| 15 | d | a | 35 | d | c |
| 16 | a | b | 36 | c | a |
| 17 | d | d | 37 | b | a |
| 18 | d | b | 38 | c | b |
| 19 | c | c | 39 | a | b |
| 20 | d | a | 40 | b | a |